Beautiful Antique Clocks from Around the World

Descriptions, Stories, and the History of these Beautiful Clocks

Curious Griffin Clock — Water-Clock — Frankfort Clock — James I.'s Clockmakers — Clock a Surname — Dutch Musical Clock — Foreign Clockmakers in London — Charles I. charters the Clockmakers' Company — Early History of the Company — Clock of St. Margaret's, Westminster — Clock at Augsburg — Clock in London Pageant — Clock in a Jesuits' College — Venice Clock — Streets named from Clocks — Clock which regulated the time of Charles I.'s Execution — Oldest Clock in America — Oliver Cromwell's Clock — Charles II.'s Gift of a Clock — Clock in Kendal Museum — Evelyn's Entries about Clocks — Ball or Bullet Clocks — Pepys sees curious Clocks — Elementary, or Earth, Air, Fire, and Water Clocks — Fulham Clock — Sir Edward Lake's Gift of a Clock — Alarum Clock mentioned in a State Paper — Pepys sees a curious Clockwork — Charles II.'s Clocks — Cosmo III.'s Record of a Clock — St. Dunstan's Clock — Its Mechanical Figures — Clock at Bristol — Bennett's Mechanical Clock Figures — Sir Matthew Hale's Clock — Milton's Lines for a Clock Case — Dryden's Reference to a Clock — Duchess of Gloucester's Striking Clock — Clock by Tompion — Clock at Windsor Castle — Clockmakers' Tokens — Grollier's Clocks — Inclined Plane Clocks — Clock mentioned in a Satirical Poem — Derham's 'Artificial Clockmaker' — Mechanical Clock made in compliment to Louis XIV. — Versailles Clock — Maria Antoinette's Clock — Clock of the King's Death — Clock at St. Cloud.

At a meeting of the Archæological Institute, held on March 1st, 1861, and also at a meeting of the Society of Antiquaries, held on June 20th in the same year, Mr. O. Morgan exhibited a miniature clock in the form of a square tower surmounted by a dome, on which stood the figure of a boy playing on a lute. The height of the clock without the dome was only one inch and three quarters. The case was of silver gilt, the works of steel. It went twelve hours, struck, and had an alarum. Mr. Morgan believed it to be of German work, and placed its date about the year 1600. It was the smallest standing clock he had ever met with. And at a meeting of the same institute, held on December 7th, 1855, the same gentleman exhibited two clocks of novel design and construction. One was in the form of a griffin, bearing an escutcheon, on which was the dial. The animal constantly rolled its eyes whilst the mechanism was in movement, and it opened its mouth when the quarters struck, and flapped its wings at the striking of the hour. The other was in the form of a crucifix; the hours were shown on a globe which revolved on the top of the cross. The date of these strangely-shaped clocks was the early part of the seventeenth century. In the South Kensington Museum is another clock arranged as a crucifix. It is of ebony, silver, and gilt and enamelled bronze. It is of French or Flemish manufacture, of the seventeenth century. Its height is fourteen inches and a half. It was purchased for 6*l*.

The Hon. W. B. Warren Vernon has a clock of gilt metal in the form of a turret, with a pinnacle top, upon a stand of ebony. It has chased silver ornaments, and is of the seventeenth century, and of French manufacture. In the South Kensington Museum is a medallion clock, in a rock-crystal case, supported on a baluster-shaped crystal stem. It is dated 1609. Its height is seven inches and three quarters. It was purchased at the Bernal sale for 14*l*. Mr. A. J. B. Beresford Hope has a gilt metal clock with columns and pinnacles at the angles, engraved dial, and square *repoussé* stand. It was made at Strasburg in 1614. Also a gilt metal table-clock of hexagonal shape, with glass panels, resting on six terminal figures. It is of the seventeenth century.

In a manuscript of the beginning of the seventeenth century, preserved in the British Museum, and entitled

'A Catalogue of Natural and Artificial Curiosities within the County of Oxford,' mention is made of "a clock that goes by water at Hanwell."

In 1605 a clock was put up in the cathedral at Frankfort. It consisted of three parts or divisions; in the lowest, which looked like a calendar, were several circles, the first of which showed the days and months; the second, the golden number, with the age and change of the moon; and the third the dominical letter. The fourth and fifth circles represented the ancient Roman calendar. On the sixth were the names of the apostles and martyrs, the length of the days and nights, and the entrance of the sun into the twelve signs of the zodiac. The seventh and eighth circles exhibited the hours and minutes when the sun rose and set. In another circle the divisions of the twelve signs of the zodiac, the four seasons, and the twelve months were marked. A circle in the centre showed the movable feasts. The figures which struck the hours represented two smiths with hammers in their hands. This piece of mechanism was repaired for the first time in 1704.

In Devon's 'Issues of the Exchequer,' the Pell Records of the time of James I., we find the following entries:—1605, 10th of October. "By Order, the last of September, 1605. To Uldrich Henche, clockmaker, or to his assignee, the sum of 100*l*. for a clock, in manner of a branch, made by him, and set up in his Highness's chamber at Whitehall. By writ dated 23rd of July, 1605. 100*l*."—1607, 5th of July. "To Humphrey Flood, goldsmith, or his assigns, the sum of 120*l*., in full satisfaction and payment for a clock covered with gold and set with diamonds and rubies, and by him delivered to his Majesty's use, at the price of 220*l*., whereof received 100*l*. By a Privy Seal dated the 5th of July, 1607. 120*l.*"—1617, 1st of April. "By Order, dated 29th of March, 1617. To Ranulph Bull, keeper of his Majesty's great clock, in his Majesty's palace of Westminster, the sum of 56*l.* 13*s.* 4*d.*, in full satisfaction and discharge of and for divers sums by him disbursed for mending the said clock, in taking the same and other quarter clocks all in pieces, and repairing the same in the wheels, pulleys, hammers, weights, and in all other parts, and in new hanging, wiring, and cording of the same clock, and other necessary reparations thereunto belonging, the charge whereof, with his own workmanship and travail therein, doth amount to the sum aforesaid, appearing by a note of the particular demands, delivered upon his oath, taken before one of the Barons of his Majesty's Exchequer, without account or imprest to be made thereof. By writ, dated 27th of March, 1617. 56*l.* 13*s.* 4*d.*" It will be observed that the last payment was made to Ranulph Bull. We find in an account of the household expenses of Prince Henry, in 1610, Emanuel Bull, the "clocke-keeper," mentioned. Probably he was some relation of the clock-keeper of James I., if not the same identical person.

In the collection of State Papers of the year 1622, in the same king's reign, there is a warrant to pay 232*l.* 15*s.* to David Ramsey, the king's clockmaker, for repairing clocks at Theobalds, Oatlands, and Westminster, and for making a chime of bells adjoining to the clock at Theobalds. On the accession of Charles I., Ramsey was again appointed the king's clockmaker, and was also nominated the first Master of the Clockmakers' Company in the charter of incorporation in 1631.

Among the curiosities of horology is the fact that the word clock has become a surname. Thus, we find in the 'Calendar of State Papers,' *temp*. James I., under date May 1st, 1609, a power-of-attorney from Sir Edward Musgrave, of Hayton, Cumberland, to John Clock, of Staple Inn, Middlesex, to receive on his behalf 20*l*., lent by him to the king on privy seal, on July 31st, 1604. And in 1618 a man named Peter Clocke was living in St. Olave's, Southwark. He is thus entered in a return now in the State Paper Office:—"Petter Clocke; b. in Turnney; in England 4 yeres; and his wyfe, taffety wever; 2 children borne in Tornney." The 'Annual Register' for 1775, tells us of letters patent granted to William Clockworthy, for the sole use of a discovery of certain materials for the making of porcelain.

Among the State Papers of the time of James I. there is an original letter, dated August 4th, 1609, addressed by Sir Julius Cæsar to the clerks of the Signet, requesting them to prepare a warrant to pay 300*l*. to Hans Niloe, a Dutchman, for a clock with music and motions. And on the 17th of the same month Sir Julius wrote from the Strand to Salisbury, stating that he was pressed by Hans Niloe for the 300*l*. for his clock.

It would seem that foreign artizans were at this period extensively employed in clockmaking in England; and it appears from documents which are now in the State Paper Office, that early in the seventeenth century there were in London fifteen clockmakers and two watchmakers, all of whom were foreigners. By "A true Certificat of the Names of the Straungers residing and dwellinge within the City of London," &c., taken by direction of the Privy Council, by letters dated September 7th, 1618, we find that in the ward of Farringdon Within was then living "Barnaby Martinot, clockmaker; b. in Paris; a Roman Catholicque." In Portsoken ward was living "John Goddard, clockmaker; lodger and servant with Isack Sunes in Houndsditch; b. at Paris, in Fraunce; heer 3 yeers; a papist; yet hee hath the oath of allegiance to the king's supremacy, and doth acknowledg the king for his soveraigne dureing his abode in England; and is of the Romish church."

In the year 1622, the clockmakers of London complained to James I. of the great number and deceitful tricks of foreigners practising their trade, and begging that they might not be permitted to work except under English masters, and that no foreign clocks might be imported. In 1631 the Clockmakers' Company of London was incorporated by a royal charter, dated August 22nd, in that year, from Charles I., as "The Master, Wardens, and Fellowship of the Art of Clockmakers of the City of London." The company had power by their charter to make by-laws for the government of all persons using the trade in London, or within ten miles thereof, and for the regulation of the manner in which the trade should be carried on throughout the realm. And in order to prevent the public from being injured by persons "making, buying, selling, transporting, and importing any bad, deceitful, or insufficient clocks, watches, larums, sundials, boxes, or cases for the said trade," powers were given to the company by the charter "to enter with a constable or other officer any ships, vessels, warehouses, shops, or other places where they shall suspect such bad and deceitful works to be made or kept, for the purpose of searching for them;" and if entrance should be denied they might effect it by force. This right of search was constantly exercised until the end of the century. The town was divided into districts, periodical searches were made, and many instances are recorded of deceitful works being found and broken up.

By the charter of incorporation of the Clockmakers' Company, David Ramsey was elected the first Master, and Henry Archer, John Wellowe, and Sampson Shelton were the first wardens. The Company, from their establishment in 1631, having no hall, held their meetings at some tavern in the city. Their last meeting before the fire was held on August 20th, at the 'Castle' tavern, in Fleet Street; and the first meeting after, on October 8th, 1666, at the 'Crown' tavern, in Smithfield. After the fire the 'Castle' tavern was rebuilt; and in October, 1735, the obituary of that month records the death of Sir John Tash, Knight, Alderman of Walbrook ward, who formerly kept the same tavern, and was one of the most considerable wine-merchants of London. He was then in the sixty-first year of his age, and commonly reported to be worth 200,000*l.* The Company tried very hard for a livery for many years, and upon one occasion when they petitioned for it to the Court of Aldermen, there was a very long discussion thereon, and upon a division the twenty Aldermen present voted as follows: for, six; against, twelve; neuter, two, consequently the petition was again rejected. This was on July 11th, 1749; but a few years later the Company succeeded; and on Lord Mayor's Day, November 9th, 1767, at the inauguration of the Right Honourable Thomas Harley, brother to the Earl of Oxford, as Chief Magistrate, we are told in a newspaper of the following day, that "The Tin-plate Workers' Company and the Clockmakers' Company joined the Lord Mayor's procession yesterday for the first time since they were made Livery-Companies, and made a genteel appearance."

The Company is strictly a trade one, and is governed by a Master, three Wardens, and twenty-eight Assistants, chosen out of the fellowship. They have no hall, their meetings being held at the 'London Tavern,' where they have several paintings which were presented to them by old members.

The Company has a lending library, rich in English and foreign works on horology and the allied sciences, with a printed catalogue, published in 1830; and a cabinet of specimens of watches, containing many rarities; also of watch and clockwork, illustrative of the progress of horology from its commencement. The Company's motto appropriately runs "Tempus Rerum Imperator" (Time the Governor of all things).

In the year 1698 an Act was passed, 9 William III., c. 28, for protecting both the trade and the public against certain fraudulent impositions then practised by foreigners and others. The second clause of the statute recited that "great quantities of boxes, cases, and dial-plates for clocks and watches have been exported without their movements, and in foreign parts made up with bad movements, and thereon some London watchmakers' names engraven, and so are sold abroad for English work; and also there hath been the like ill practice in England by divers persons, as well as by some professing the art of clock and watchmaking, as others ignorant therein, in putting counterfeit names, as also the names of the best known London watchmakers on their bad clocks and watches, to the great prejudice of the buyers, and the disreputation of the said art at home and abroad." For the preventing of such ill practices it was ordained that no case or dial-plate should be in future exported without the movements, nor without the maker's name and place of abode engraven on every such clock and watch, under a penalty of forfeiture and a fine of twenty pounds. In the 27th year of the reign of George II., c. 7, another clock and watchmakers' act was passed.

The journal of the Clockmakers' Company makes mention of a clock which cost 3*l.* in 1636.

In the churchwardens' accounts for St. Margaret's, Westminster, under date 1617, is an "Item, paid to Leonard Tenant, for a new clock and chimes and twoe dyals, and for a barrel, and pricking thereof, and for wyers to the chimes, and for all the iron-work, and workmanship in the setting up, and finishing of the same, according to an agreement made with him in that behalf, being for the use of this church and parish, 37*l.*" And under date 1658 is an "Item, to Mr. Farmer, for making of the new diall on the west-end of the church, as by his bill appeareth, 14*l.* 10*s.*;" and "Item, to Mr. Farmer, for a new dyall at the west end of the church on the churchyard side, 7*l.*"

In the 'Monasteriologia' of Stengelius, published in 1619, is an engraving of the monastery of Saints Ulric and Afra, at Augsburg, showing a clock, with a bell in a turret over it, in the upper part of a building adjoining the gateway entrance.

An account of the London Pageant given in honour of James I., in 1620, tells us that a part of the show "was a chariot painted full with houre-glasses and sundialls; the fore-wheeles were two globes, and the hinder wheeles were like two church-dialls. Within it aged Time was drawne, seated upon an houre-glasse."

In the State Paper Office is an inventory of goods which were found in a Jesuits' College in Clerkenwell, in 1627–8; and among the articles enumerated therein are "one strikeinge clocke" and "one hanginge watch with an alarum."

In the South Kensington Museum is a table-clock in gilt bronze, the sides of which are decorated with allegorical figures in relief, representing impersonations of Arithmetic, Geometry, Music, and Astrology. It is of German work, about the year 1640. It was purchased for 6*l.*

So early as the seventeenth century there was in a tower in the Grand Piazza at Venice, an old clock, which not only pointed out the hours and their subdivisions, but also exhibited the signs of the zodiac, with the courses of the sun and moon. On certain festivals, and especially every hour while Ascension fair lasted, the statues of an angel and the three kings, or eastern magi, made their appearance at a door in this clock, and in passing made obeisance to the figures of the Virgin and Child placed in a niche, and then returned through another door on the opposite side. On the top of the tower were two brazen Moors, or, as Longfellow calls them, "bronze giants," who struck the hours with hammers on a large bell. This piece of clockwork resembled that at Macerata; but the images in the former were larger.

Evelyn, in his 'Memoirs,' under date 1645, records that while he was at Venice he went "thro' an arch into the famous Piazza of St. Marc. Over this porch stands that admirable clock, celebrated next to that of Strasburg for its many movements; amongst which, about 12 and 6, which are their houres of Ave Maria when all the towne are on their knees, come forth the 3 Kings led by a starr, and passing by ye image of Christ in his Mother's armes do their reverence, and enter into ye clock by another doore. At the top of this turret another automaton strikes ye quarters. An honest merchant told me that one day walking in the Piazza, he saw the fellow who kept the clock struck with this hammer so forceably, as he was stooping his head neere the bell to mend something amisse at the instant of striking, that being stunn'd he reel'd over the battlements and broke his neck."

This clock was thus described early in the present century, when it was seen by the writer:—"The Horological Tower is in the splendid square of the Imperial Palace at Venice. It is also called the tower of the clock; it contains the city clock and a bell, with two large bronze human figures, who, with huge hammers, regularly strike the hours. Midway up the Horological Tower sits a noble bronze gilt figure of the Virgin and the infant Jesus, with an open gallery in front of her, facing the square. On each side is a door opening into the interior. At the striking of the clock, these doors fly open, and several persons move out in succession; the first is a trumpeter, who raises his trumpet to his mouth as he comes in front of the Virgin; then follow three others in succession, dressed like eastern sages, and one of them a person of colour. They all pass in front of the Virgin round to the other side, bowing as they pass; then halting a moment, they straighten up, and entering the other door disappear. This is called the Visit of the Magi."

White, in his 'Fragments of Italy and the Rhineland,' 1841, tells us that "At a certain period of every year, namely, on the Feast of the Ascension, and fourteen days afterwards, the great clock tower of St. Mark's, which by the way is an edifice boasting a barbaric confusion of ornament that rivals any of its fellows in that Noah's ark of architecture, exhibits a curious piece of mechanism. On the summit of the Orologio is a great iron bell, forming a kind of cupola to the gateway, and heaving its black form against the bare sky. On each side stands a savage man of Ind, grasping an iron mace, with which he beats the hour upon the bell. On a platform midway in the building, and just over the huge gold and blue enamelled disk of the dial, is a balcony of gilt lattice, surrounding an image of the Blessed Virgin and Child, who is seated between two doors all overlaid with gold. Whenever the clock strikes, the door on her right hand opens, and an angel with a trumpet in his hand steps out, he acts as gentleman-usher to the Three Royal Magi, who pass before her, just as you might do before Queen Victoria at a levee, only that, instead of kissing hands, each raises his diadem or turban from his brow, makes a graceful obeisance, and in stately procession re-enters the turret by the other door. The figures are apparently as large as life, and gorgeously apparelled; and it really is a pretty pageant."

A street in Venice is named after this Orologio—a kind of nomenclature of which we have another example at Rouen, where the Rue de la Grosse Horloge, or the street of the great clock, is so called from a huge, clumsy, antiquated clock, that is attached to the upper part of an arch across the street.

At Vicenza, in the Piazza de' Signori, which is a minified copy of the Piazza San Marco at Venice, is a small Torre dell' Orologio, or clock-tower, somewhat after the style of the one at the latter city, above described.

The 'Gentleman's Magazine' for 1799 informs us of the death at Green Street, on March 24th, in that year, of Mrs. Forester, aged eighty-three, the widow of the then late Rev. Dr. F. Moore, and the daughter of the Hon. and Rev. Dr. Moore. This lady possessed at Great Brickhill, Bucks, the identical clock which was at Whitehall at the time of the execution of Charles I., in the year 1649, and by which the fatal moment was regulated. The effects of Mrs. Forester were shortly after her death dispersed by public sale.

The Philadelphia Library claims possession of the oldest clock in America; it wants but a few years of being two centuries old. It was made in London, keeps good time, and is said to have been once owned by Oliver Cromwell.

Charles II., according to tradition, gave to Mrs. Jane Lane a clock in memory of her services after the battle of Worcester. On the clock was engraved the name of "Henricus Jones, Londini." In North's 'Life' it is stated that barometers were first made and sold by one Jones, a noted clockmaker, in the Inner Temple Gate, at the instance of the Lord Keeper Guildford. Probably Jones was the first Englishman who constructed a Torricellian tube, as the barometer was originally called after its inventor, Evangelista Torricelli, who between the years 1641 and 1647 discovered the means of ascertaining the weight of the atmosphere by a proportionate column of quicksilver.

In the Museum of Kendal is an interesting specimen of a clock, which is said to be one of the first ever made on the pendulum principle. It stands on a bracket; it has one weight suspended by a cord, and only one hand; it goes twelve hours; but by cutting away the floor beneath it, it could be made to go twenty-four. The bell forms a dome over the clock; it has an excellent sound, and strikes all the hours. On each side of the case is a brass door. The clock still keeps correct time. On the dial is the following inscription: "J. C. George Pooll in S. Ans Lane Fecit. The Guift of James Cock, Maior of Kendall, 1654, to the Maior of the same sucksesiuely. Time runeth; your work is before you." Old Humphrey wrote one of his 'Pithy Papers' upon the subject of this clock and its motto, and has thereon hung a moral.

Evelyn in his 'Diary' has three curious entries respecting clocks. Under date February 24th, 1655, he says: "I was shew'd a table clock whose ballance was onely a chrystall ball sliding on parallel wyers without being at all fixed, but rolling from stage to stage till falling on a spring conceal'd from sight, it was throwne up to the upmost channel againe, made with an imperceptible declivity, in this continual vicissitude of motion prettily entertaining the eye every halfe minute, and the next halfe giving progress to the hand that shew'd the houre, and giving notice by a small bell, so as in 120 halfe minutes, or periods of the bullet's falling on the ejaculatorie spring, the clock part struck. This very extraordinary piece (richly adorn'd) had been presented by some German Prince to our late King, and was now in possession of the Usurper, valu'd at 200*l*." Under date November 1st, 1660, Evelyn says, "I went with some of my relations to Court, to shew them his Maties cabinet and closset of rarities. . . . . Here I saw . . . . amongst the clocks, one that shew'd the rising and setting of the Sun in ye Zodiaq, the Sunn represented by a face and raies of gold, upon an azure skie, observing ye diurnal and annual motion, rising and setting behind a landscape of hills, the work of our famous Fromantel." On August 9th, 1661, Evelyn says, "I din'd at Mr. Palmer's in Gray's Inn, whose curiosity excell'd in clocks and pendules, especialy that had innumerable motions, and plaied 9 or 10 tunes on the bells very finely, some of them set in parts, which was very harmonious. It was wound up once in a quarter."

Pepys in his 'Diary,' under date July 28th, 1660, records that he went "To Westminster, and there met Mr. Henson, who had formerly had the brave clock that went with bullets (which is now taken away from him by the King, it being his goods)." After the dates given by Evelyn and Pepys as above it was not uncommon for clocks to be made with a small ball or bullet on an inclined plane, which dropped every minute. Gainsborough, the painter, had a brother who was a dissenting minister at Henley-on-Thames, and who possessed a strong genius for mechanics. He invented a clock of very peculiar construction, which after his death was deposited in the British Museum. It told the hour by a little bell, and was kept in motion by a leaden bullet, which dropped from a spiral reservoir at the top of the clock into a little ivory bucket. This was contrived so as to discharge it at the bottom, and by means of a counter-weight it was carried up to the top of the clock, where it received another bullet, which in its turn was discharged like the former. This seems to have been one of the many attempts made at perpetual motion.

In 1663, Martinelli, of Spoleto, wrote a curious work, describing various methods of constructing what he calls elementary clocks, that is, clocks which were set going by earth, air, fire, and water; some of which could be made to show the time of day, the days of the week and month, the courses of the moon and planets, with the Epact. Time was measured in the water-clocks by suffering that element to pass successively through the compartments of a drum-shaped cylinder, acting as a pulley to a cord with a counter-weight, the rapidity of the motion being determined by the quantity of the water, or the bore of the orifice through which it escaped. The motion of the earth, or sand clock, was regulated in a similar manner. In the air-clock time was measured by the pumping of a bellows, like those of an organ, the gradual escape of the air regulating the descent of a weight, which carried round the wheels, as in other time-keepers. In the fire-clock the motion was produced upon the principle of a modern smoke-jack, the wheels being moved by means of a lamp, which also gave light to the dial; and the clock could be made to announce the several hours by placing at each a corresponding number of crackers, which by certain contrivances were exploded at proper times. He tells us that these clocks offered considerable advantages to persons troubled with insomnia, or want of sleep, as they gave a soft light, and without noise marked the silent flight of time.

In Bagford's Collections in the British Museum, Harl. MSS. 5931, is the following advertisement of a clock which was moved merely by the exhalations of a lighted candle:—"The Turkish Seraglio, in Waxwork. The Story of Queen Voadicia, &c. The Temple of Ephesus, &c.; and of Apollo; the vision of Augustus; and the six Sibyls, &c., the fatal sisters, that spin, reel, and cut the thread of man's life, &c. Moving figures, &c. An old woman flying from Time, who shakes his head and hour glass with sorrow, at seeing age so unwilling to die. Nothing but life can exceed the motion of the heads, hands, eyes, &c., of these figures, &c. Other curious pieces of clockwork, and rarities, &c. A clock like the long pendulums now in use, but internally different, the motions of the two hands and striking regulated without pendulum or balance or fly, by the exhalations of a lighted candle, not hid, but exposed, and of the same use as if placed in a candlestick, &c. Mrs. Salmon teaches the full art, sells all sorts of moulds and glass eyes, with other materials, and takes likenesses of gentlemen and ladies, in St. Martin's, near Aldersgate-street. Prices *6d.*, *4d.*, and *2d.*" Mrs. Salmon removed soon after from St. Martin's-le-Grand to the Golden Salmon, at Temple Bar, which she says was a more convenient place for the coaches of the quality to stand unmolested. Her collection of waxwork figures was for half-a-century one of the most popular exhibitions in London, especially to country visitors.

We read in Faulkner's 'History of Fulham and Hammersmith,' that the clock in Fulham Church steeple was the gift of an individual in order to exonerate himself from serving any office during his residence in the parish, as appears by the following extract from the parish books, dated August 14th, 1664:—

"Ordered, that Richard Goslinge, of this parish, brickmaker, bee and is from this day forward, during his abode in this parish, quitted from bearing any office off and belonging to the parish of Fulham, upon condition if the said Richard Goslinge doe, at his own proper costs and charges, give an able and substantial clock, not under the value of *12l.*, and y[t] the old clock bee given unto the said Rich[d]. Goslinge, which new clock is the voluntary gift of him the said Rich[d]. Goslinge, in consideration of the privileges aforesaid."

Sir Edward Lake, by his will dated April 8th, 1665, gave as follows:—"To the church or chappell of Normanton, near Pontefract, in Yorkshire (if there be a church or chappell there, which I know not), where my paternall ancestors have lived for many years," a clock, and a sum "for the maintaining and keeping of it for ever." In Normanton Church is this inscription referring to the gift:—

> "Edwardus Lake de Norton Episcopo, in
> Comitatu Lincolniensi Eques Auratus
> LL.D. Dioceseos Lincolniensis Cancellarius
> In majorum memoriam qui olim in hoc oppido
> Normantoniæ habitaverunt, hoc Horologium
> Dedit, ac etiam decem solidos ad reparationem
> Ejusdem annuatio in perpetuam solvendos
> Deo et Carolis Regibus dominis suis
> Presertim, Carolo Martyri Pacis et Belli
> Tempore fideliter nec non insigniter inserviit.
> Et animam Deo pie reddidit 18 die Julii anno
> Etatis suæ 77 annoq. Domini 1674.
> Et Ecclesia Cathedrali Beatæ Mariæ Lincoln.
> Sepultus jacet."

In the 'Calendar of State Papers,' *temp.* Charles II., under date May 10th, 1666, we find a letter in French from M. de Marainville to M. de la Fabvollière, requesting him either to return to M. De Samborne the alarum-clock which he, the writer, had lent him, or if it suited him, and he wished to keep it, to pay for it seven pieces, the price the clockmaker at Charing Cross offered for it.

On September 6th, 1667, the sight-loving Pepys went with his wife to Bartholomew Fair, and after witnessing a man whose legs were tied behind him dance upon his hands, next "went to see a piece of clockework made by an Englishman—indeed, very good—wherein all the several states of man's age, to 100 years old, is shewn very pretty and solemne."

In the Camden Society's 'Secret Services of Charles II. and James II.,' vol. lii., are various accounts of payments made on behalf of the King, some of which were for clocks supplied to him. Thus, in the account up to

April 3rd, 1668, is an item, paid "To Humfry Adamson, for a clock by him sold for the chappell at Whitehall, 19*l.* 7*s.*" In the account up to March 9th, 1682, is an item, paid "To Wm Chiffinch, for so much money he paid Saml Watson, for a clock he sold his late Ma'tie, wch showes the rising and setting of the sun and moon, and many other motions, 215*l.*" His late majesty must have been Charles I. In the account up to July 3rd, 1682, is an item, paid "To Mr. Knibb (no doubt the same person as the one to whom we have referred at page 112), by his said Ma'ties comand, upon a bill for clockwork, 141*l.*" In the account up to December 9th, 1682, is a debit, "To Robt Seignior, for a clock bought of him, and sett up in the Trea'ry Chambers, for the use of the Commissioners of his said Mat^{s} Trea'ry, 20*l.*" All these payments were made in the reign of Charles II.

In 1669, Cosmo III., Grand Duke of Tuscany, relates in his travels that he saw at the Royal Society of London, a clock, the movements of which were derived from the vicinity of a loadstone; and it was so adjusted as to discover the distance of countries at sea by the longitude. This clock, says Mr. Weld, in his 'History of the Royal Society,' together with Hooke's magnetic watches, were on the occasion of the visit of illustrious strangers, always exhibited as great curiosities.

The mechanism of Venice clock, which we have before described, was in one particular imitated in the clock of St. Dunstan's Church, Fleet Street, which was taken down in 1831. This remarkable clock, that projected over the street in the manner of those of several of the city churches at the present time, was set up in the year 1671. The artist, Thomas Harrys or Harris, received for his work the sum of 35*l.*, and the old clock. It appears by the parish-books, that on May 18th, 1671, Thomas Harrys, who was then living at the lower end of Water Lane, London, made an offer to build a new clock with chimes, and to erect two figures of men with pole-axes to strike the quarters. This clock was so constructed as to afford one dial plate at the south front, and another at the east end of the church. All this he proposed to perform, and to keep the whole in constant repair for the sum of 80*l.* and the old clock; at the same time observing that his work should be worth a hundred pounds. He further adds these words: "I will do one thing more which London shall not show the like; I will make two hands show the hours and minutes without the church, upon a double dial, which will be worth your observation, and to my credit." It appears that the vestry agreed to give to Harrys the sum of 35*l.* and the old clock for as much of his plan as they thought proper to adopt; and on October 28th, in the same year, 1671, his task being completed, he was voted the sum of 4*l.* per annum to keep it in repair. We find that the idea of chimes was given up, as well as the dials at the east-end. Originally, in 1737, this clock, with its large gilt dial, was within a square, ornamented case, with a semi-circular pediment, and the tube from the church to the dial was supported by a carved figure of Time, with expanded wings as a bracket. In 1738 it cost the parish 110*l.* for repairs. Above it in an alcove, and in a standing posture, were two life-size wooden figures of "savages or Hercules," as Strype describes them, or "two wooden horologists," as Ned Ward calls them, with clubs in their right hands, who struck the quarters of every hour on the two suspended bells, moving their heads at the same time. These figures much excited the interest of the passers-by, especially provincial visitors to London, who would stop in crowds to see these automata strike the quarters with their clubs. Leigh, in his 'New Picture of London,' calls them the "pets of cockneys and countrymen:"—

> "Many a stranger as he passed that way
> Made it once a design there to stay
> And see those two hammer the hours away
> In Fleet Street."

They were one of the sights of London, and many a gazer at them has unwittingly enriched the pickpockets and cutpurses who used to mix with the crowd of gaping idlers assembled under this clock, to the no small obstruction of the foot and carriage-way. One historian tells us that they were "more admired by many of the populace on Sundays than the most eloquent preacher from the pulpit within." A writer in the 'Mirror,' 1828, says, "It would be needless to describe the two brazen striking Saracens who attract the gaping multitude; when they perform operations one would really suppose they were in league with the pickpockets, who are below striking into the pockets of their admirers sans cérémonie." The author of 'London Scenes and London People,' an eye-witness of the old clock, says, "The giants stood in front of the building, about 30 feet from the road, on a covered platform, each wielding a club—the bell being hung between them, which at the quarters, as well as whole hours, they struck, but so indolently, that spectators often complained that they were not well up to their work. The mechanism, too, was rough and clumsy; you could not help noticing the metal cord inserted in the club, to which its motion was due." Sir Walter Scott speaks of the savages in his 'Fortunes of Nigel;' but he places them in position before they were known to the gaping cockneys. Cowper thus alludes, in his 'Table Talk,' to these figures :—

> "When labour and when dullness, club in hand,
> Like the two figures at St. Dunstan's, stand
> Beating alternately, in measured time,
> The clockwork tintinnabulum of rhyme,
> Exact and regular sounds will be;
> But such mere quarter strokes are not for me."

In 'A Pacquet from Wells; or, a New Collection of Original Letters,' &c., 1701, we read: "A Lady of Pleasure being the Escutcheon of Iniquity, and the Cully and Pully her two Supporters, hanging thus like St. Dunstan's Clock, between Boucher and Bowden for both to knock at in their turns." When the old church was pulled down the clock and figures were purchased by the Marquis of Hertford, and removed to his villa in Regent's Park, where the clubbers still do duty every quarter of an hour. We read under date October 22nd, 1830, "Mr. Creaton, auctioneer, sold by private contract to the Marquis of Hertford the clock-tower, with its two figures, for 210*l.*"

St. Dunstan's had a clock previous to 1671. It was an overhanging one, and beneath it, and in front of the church, were shops inhabited by booksellers. One of the titles to the publications issued from here reads: "Celia, containing certaine Sonets. By David Murray, Scoto Britaine, at London, Printed for John Smethwicke, and are to be sold at his shop in St. Dunstan's Church-yard, in Fleet Street, *under the Diall*, 1611," 12mo. What made the "Diall" so famous was the setting up of the giants in 1671; although so early as 1478 there was a similar piece of mechanism in Fleet Street. Stow describes a conduit erected that year near Shoe

Lane, with angels having "sweet sounding bells before them; whereupon, by an engine placed in the tower, they divers hours of the day and night with hammers chimed such an hymn as was appointed." Mr. Denham tells us that "whatever St. Dunstan's clock might have been in the first instance, it is certainly not the only device of the kind extant at the present time in England. Those persons who have inspected the curiosities at the Cathedral at Norwich will remember that the quarters for the use of persons within the building are struck by two similar, though much smaller figures, placed near two bells, inside the church in one of the recesses of the south aisle, and that the arms communicate with the abbey clock by strings, which are visible in their whole course from the figures to the ceiling." Old St. Paul's Cathedral, London, had automaton figures which struck the quarters on the clock bell, as we have before stated.

The old church dedicated to the Holy Trinity at Bristol, which was demolished in 1787, the new one of Christ Church occupying its site, had a tapering spire, and in the tower was a clock guarded by gigantic "quarter-boys," represented in two large figures, with ever-ready hammers to note the flight of time. They were similar to those formerly outside St. Dunstan's Church; and were placed under a semi-circular canopy on each side of the face of the clock. They wore brass helmets, and were partly habited in armour; each grasped a battle-axe, with which it struck the bell suspended over its head. It would appear that they were coloured and gilt with great care, according to the taste of the age. These "quarter-boys" are still preserved at Bromfield House, Brislington, in the possession of the Rev. George Weare Braikenridge.

Mechanical figures similar to those which stood at St. Dunstan's Church have lately been set up at the shop of Bennett, a watch and clockmaker, No. 65, Cheapside. There are two recesses in front of the premises, and in the upper one a figure of Time has the double duty of carrying a scythe and hour-glass, and every sixty minutes striking a large and sonorous bell. In the lower recess figures of Gog and Magog, modelled by Brugiotti from the originals in Guildhall, strike the quarters of the hour. In front of the latter recess is a large projecting clock, the illuminated dials of which make it useful by night as well as by day. Above the parapet of the house-top, about twenty feet, a ball, fifteen feet in circumference, falls every hour by an electric current transmitted from the Observatory at Greenwich.

Clocks occupy a very high place among the instruments by means of which human time is economized, and their multiplication in conspicuous places in large towns is attended with many advantages. Their position, however, in London is often very ill chosen; and the usual place, half-way up on a high steeple, in the midst of narrow and crowded streets, is very unfavourable, unless the church happen to stand out from the houses. The most convenient situation for a clock is a projecting elevation above the street, with a dial-plate on each side, like that of old St. Dunstan's Church, and that of Bow Church, Cheapside, at the present time.

The clock at Alderley is a very old and curious piece of mechanism. An inscription on it states that it was presented to the church of Alderley by the great Sir Matthew Hale:—"This is the Guift of the Right Honourable the Lord Chief Justice Heale, to the

Parish Church of Alderly. John Mason, Bristol, Fecit, Novem. 1st, 1673." It appears by this inscription to have been presented on his birthday, which from the record on his tomb was November 1st. Alderley is the family place of the Hale family to this day.

The following beautiful poem, to which no date is assigned, was written by Milton, who was born in 1608, and died in 1674; therefore probably it was composed about the period of which we are writing. Before these verses, it appears from the author's manuscript, he had written, "To be set on a clock-case."

"On Time.

"Fly, envious Time, till thou run out thy race;
Call on the lazy leaden-stepping hours,
Whose speed is but the heavy plummet's pace;
And glut thyself with what thy womb devours,
Which is no more than what is false and vain,
And merely mortal dross;
So little is our loss,
So little is thy gain!
For when as each thing bad thou hast intomb'd,
And last of all thy greedy self consum'd,
Then long Eternity shall greet our bliss
With an individual kiss;
And joy shall overtake us as a flood,
When everything that is sincerely good
And perfectly divine,
With Truth, and Peace, and Love, shall ever shine
About the supreme throne
Of Him, to whose happy-making sight alone
When once our heavenly-guided soul shall climb;
Then, all this earthy grossness quit,
Attir'd with stars, we shall for ever sit,
Triumphing over Death, and Chance, and thee,
O Time!"

Time and time-measurers have often furnished themes to poets and moralists, as we shall have occasion to show in order of date. The end and purpose of clocks and watches are pre-eminently suggestive of things which relate to our inner life, measuring as they do so certainly and so continuously our steps from our cradles to our graves, our hours of joy and our hours of sorrow. Probably no mechanisms have more inspired and obtained a literature of their own than time-measurers, which have in all ages since their invention offered an almost inexhaustible theme for poets and moralists. The mottoes that have appeared on old sundials—those antique Altars of Time—would fill a small volume. One other example, which we may fit in here, is given by Dryden, who was born in 1631, and died in 1701, and who, alluding to the death of King Polybus, says in his 'Œdipus,'

> "Till like a Clock worn out with eating Time,
> The wheels of weary Life at last stood still."

During the seventeenth century there was a great taste for striking-clocks, which were to be had in every variety of form. Several of them, made by Thomas Tompion, who invented many useful things in clock-work, not only struck the quarters on eight bells, but also the hour after each quarter. At twelve o'clock forty-four blows were struck, and one hundred and thirteen between twelve and one o'clock. Failures in the striking mechanism of these clocks were attended with much annoyance to the owners of them; for they would go on striking without cessation until the weight or spring had gone down, and they were frequently contrived to go for a month. A clock made by Tompion on this construction caused much annoyance to the Duchess of Gloucester soon after her marriage. This machine was fixed in an apartment adjoining her bedchamber; the failure took place at two o'clock in the morning, and, as the case could not be opened, the clock continued to strike until eight o'clock—a tintinnabulum which we think must have scared away the hovering Hymen from the bridal bed.

In June, 1826, a discovery was made of the *chef-d'œuvre* of Tompion, which had been so long lost. It was made for the Society for Philosophical Transactions, and was a year-going clock. A record exists which states that Tompion was at work on this clock when the great plague broke out in London; and on the day that he finished it he was attacked by the pestilence. His friends removed him to the Continent, where he died. On the dial was this inscription: "Sir James Moore caused this movement to be made with great care, anno Domini 1676, by Thomas Tompion." Tompion was paid one hundred guineas, and the clock was removed to the Society's house, and there in the confusion of the moment it was placed in the lumber room, where it lay without a case exactly a century and a half. One thing wonderful attended this discovery, namely, that all the steel pins on being cleared from dust were found to be as brilliant as ever they had been. The above statement as to Tompion's death seems to be hardly consistent with the following advertisement, which appeared in 'Mercator,' No. 79, 21-4 Nov., 1713:—"Advertisement. On the 20th instant, Mr. Tompion, noted for making all sorts of the best clocks and watches, departed this life. This is to certify to all persons, of whatever quality or distinction, that William Webster, at the 'Dial and Three Crowns,' in Exchange Alley, London, served his apprenticeship, and served as a journeyman a considerable time with the said Mr. Tompion, and by his industry and care is fully acquainted with his secrets in the said art."

At Windsor Castle is an old clock, which was made by Joseph Knibb in 1677. This artist issued the following token:—Obverse—IOSEPH KNIBB CLOCK-MAKER IN OXON (in four lines). Reverse—I. K.—A clock face and hands. It seems that clockmakers, like other tradesmen, issued their tokens in the seventeenth century. We find two further examples of this coinage still preserved:—Obverse—HENRY BORGIN—A clock face and hands. Reverse—WITHOVT BISHOPS-GATE H. M. B. Obverse—WILL BRVNSLEY AT LILLY—A clock face and hands. Reverse—HOVSE AGAINST STRAND BRIDG HIS HALFPENY.

Many curious clocks were made in the seventeenth century by Nicholas Grollier de Servière, who was born at Lyons in 1596, where he died in 1689, after an eventful life. At the age of fourteen he was a soldier, and served in the army in Italy, and lost an eye at the siege of Verciel. He was also present at the battle of Prague, where he greatly distinguished himself. His knowledge of mathematics and mechanics enabled him to render much help to his country during the memorable sieges of the time of Louis XIV. After many adventures he retired from service, and found leisure to invent and manufacture many curious clocks and ingenious toys. In one of his clocks time was measured by the descent of a ball in a metal groove, twisted round columns supporting a dome. When the ball finished its descent, its weight, lifting a detent, discharged the wheelwork, giving motion to an Archimedian screw, which raised the ball to its former position, again to descend as before. In another the ball descended in diagonal lines on an inclined plane. The means of ascent in this case were hidden from view In a third the ball was made to pass within the bodies of two serpents, which by a reciprocating motion were made to swallow the ball alternately. A compound of the motions in the two last-named clocks was adopted by a scientific gentleman about fifty years ago, and time-pieces on this principle are still called Congreve clocks after him. Grollier made several clocks for the purpose of pleasing and surprising his visitors.

Thus, a small figure of a tortoise dropped into a pewter-plate filled with water, having the hours marked on the flat edge, would float round and stop at the proper hour. If moved it would again return; and if the tail were placed at the hour it would turn round, and again point with its head. A lizard was seen ascending a pillar on which the hours of the day were marked, and pointing to the time as it advanced. The figure of a mouse was also made to move on a cornice, and point to the hours marked upon it. These contrivances were in part managed by concealed magnets, which were sufficiently strong to attract, although covered over. Grollier also contrived a method for showing the time at night, by causing the dial of a clock to revolve instead of the hands. By these means the hours and quarters were brought to an illuminated space and seen together; the quarters being differently marked, and engraven on a smaller circle than the hours. The inventor stated that these clocks were less expensive and less liable to derangement than repeating-clocks, and that none but the blind would prefer repeaters to them; it being easier to look at the one than to pull the cord of the other. Grollier made some clocks to go by their own weight, descending inclined planes, and others in grooves, forming a kind of path from the ceiling to the floor. When the clock had nearly finished its descent it was lifted off, and placed again at the highest point of its path, the hands being set to time before it was restarted. Several clocks on this principle have been projected to avoid the casualties to which main-springs and weight-lines are liable. Such a clock, invented and made by the Rev. Maurice Wheeler, was exhibited about eighty-five years ago at Don Saltero's museum of rarieties at Chelsea, which was one of the popular sights of London at the close of the last century. One of these clocks, invented by M. de Gennes, was kept in equilibrium by a weight at the end of a lever. The unwinding of the springs made the weight change its position; thus altering the centre of gravity, and causing the clock to ascend the plane.

At a meeting of the Archæological Institute, held on February 6th, 1852, Mr. Forrest exhibited a travelling or table clock, in the form of a large watch; the date being about 1690, and the maker, John Rehle, of Freiburg.

A curious satirical poem, published in 1690, and entitled 'Mundus Muliebris: or, the Ladies' Dressing Room unlocked, and her Toilette spread,' mentions, as part of a fashionable lady's paraphernalia—

> "Repeating-clocks, the hour to show
> When to the play 'tis time to go."

In 1696, William Derham, D.D., F.S.S., Canon of Windsor, and a most eminent philosopher and divine, who was born near Worcester in 1657, and died in 1735, published 'The Artificial Clockmaker, or, a Treatise of Watch and Clockwork; showing to the meanest capacities the art of calculating numbers to all sorts of movements, the way to alter clockwork to make chimes, and set them to musical notes, and to calculate and correct the motion of pendulums. Also numbers for divers movements, with the ancient and modern history of Clockwork, and many instruments, tables, and other matters never before published in any other book.' This work was republished in 1700. A fourth edition with large emendations appeared in 1734; and a new edition in 1759.

About the year 1696 Burdeau, a clever mathematician, constructed a remarkable clock in compliment to Louis XIV., whom in this ingenious work he highly flattered. On a rich throne, surrounded by all the pomp of royalty, was seated "le grand Monarque;" around him stood the Electors of the German States, and the princes and dukes of Italy. These advanced towards the king, and, after doing homage, on retiring chimed the quarters of the hour with their canes. For the kings of Europe was reserved the more dignified office of striking the hours, after having paid their respects to the king. This piece of automaton clockwork was very gratifying to the French people of the time of Louis XIV.; and many who admired it persuaded the maker to exhibit it publicly. Unfortunately for Burdeau, he advertised his intentions to do so in the newspapers, and attempted too much in order to gratify the great crowd of people which collected in consequence. Knowing the stubborn and unyielding will of William III. of England towards his sovereign, the artist determined to make William's effigy more pliant, so that when its turn came, it should make a very humble obeisance to Louis. William, thus compelled, bowed very low indeed; but at the same instant some part of the machinery snapped asunder, and threw "le grand Monarque" prostrate from his chair at the feet of the British king. The news of the accident spread in every direction as an omen; the king was informed of it, and poor Burdeau was confined in the Bastile.

In the palace of Versailles, in the Salon du Conseil, is a curious clock, that plays a chime when the hour strikes, and is set in motion by machinery, by which also sentinels are made to advance, a cock to flap his wings, Louis XIV. to come forward, and a figure of Victory or Fame to descend from the skies and crown him with a golden chaplet. The Salle des Pendules in this palace is so called from a clock in it, which shows the days of the month, the phases of the moon, the revolutions of the earth, and the motions of the planets, besides the hour, the minute, and the second of the day. We may here mention incidentally that French clock cases were formerly accounted the first in the world, and those made by Boule in the time of Louis XIV. are looked upon as curiosities of good taste and workmanship.

In 1820 appeared the following advertisement:—"Mechanical Clock. The Proprietors of an elegant and superb Mechanical Clock, 8 feet high, which was the property of Maria Antoinette, late Queen of France, have the honour to present it for the inspection of the public: this piece of mechanism represents the front view of the king's palace at Versailles (the garden side): on the commencement of striking the hour a complete and beautiful clockwork begins playing 20 different airs, after which a figure, representing the Warner of Death, makes its appearance in the centre of the balcony, at each side of which is a King's Herald in a sitting position, holding the shell of a clock in their hands, to which the figure moves alternately with a majestic step, striking the hour, and then retires; immediately after 5 doors in the balcony open, and out of the centre his Majesty Henry IV. makes his appearance by announcement of trumpets by figures at each side of the balcony, when the doors close again and leave his Majesty in a standing position, surrounded by his suite; then an angel from the top, with a crown in his hands, appears, in the midst of music by different instruments, placing the crown on the head of the monarch, with a brilliant star hovering over his head; at the conclusion of the coronation and music the trumpets recommence sounding, when his Majesty salutes the public and retires with his suite, and the doors immediately close: the dial of this incomparable piece of mechanism represents the different movements of the globe, the sun, moon, and planets, showing the leap year by the months, weeks, days, hours, and minutes, with the 4 seasons; the whole of which is regulated by the movements of this extraordinary mechanical clock. At the same place may be seen also the Cuirass of his late Majesty the Emperor Napoleon, with the Cuirass and Helmet of his son, the King of Rome; and many other new invented cuirasses. All the objects are to be sold, and are on view at No. 23, Howland Street."

In the court called the Cour de Marbre, at Versailles, was a clock that had no mechanism, and only one hand, which was placed at the precise moment of the death of the last King of France, and which did not move during the whole of his successor's reign. This custom dated from the time of Louis XIII. It was a fitting memorial for a palace, to remind the inmates of their brief tenure of office. In 1838 this "clock of the King's death" was replaced. In the imperial chateau of St. Cloud, in the Salon de la Réception de la Reine, is a clock with twelve dials, marking the hours of as many capitals of Europe.

Equation Clock — Mean Time — Clock at Warwick — Clocks in South Kensington Museum — Rochester Clock — Bishop Burnet's Gifts of Clocks — Prior's Verses on a Clock — Bishop Hall's Clock — Christopher Pinchbeck and his wonderful Clocks — Cripplegate Clock — Transit Clock — Sidereal Time — George Graham — Improvements in the Pendulum — Hogarth's Masquerade Clock — Clock Lamp — Bampton Clock — Musical Clocks — Clocks at Knole — Automaton telling the Hours — Musical and Astronomical Clock — Derby Clock — John Whitehurst — Clock Inscriptions — Marquis of Worcester's Ball Clock — Night Sand Clock — York Cathedral Clocks — John Ellicott — Horse Guards Clock — The Escapement of a Clock — Infants' Nursery Clock — Repeating-Clocks — Vocal Clock — Colonel Magniac — Clock at Royal Academy, Paris — the Microcosm — Young's Reference to Horology.

THE first Equation Clock, an ingenious contrivance to show both mean and apparent time, was made in London between one hundred and fifty and two hundred years ago. The first one invented was in the cabinet of Charles II., King of Spain, and is thus described in a letter from Father Kresa, a Jesuit, to Mr. Williamson, a watchmaker:—"I have to tell you that from the years 1699 to 1700 there has been in the cabinet of King Charles II., of glorious memory, King of Spain, a clock with a royal pendulum (seconds pendulum), made to go with weights and not with springs, going four hundred days without requiring to be once wound up . . . I had orders to go every day to the palace during several months to observe the said clock and compare it with the sundial; and at that time I remarked that it showed the equation of time equal and apparent, exactly according to the tables of Flamstead."

In the 'Philosophical Transactions,' 1719, is a letter by Joseph Williamson, a watchmaker, asserting his right to the curious invention of making clocks to keep time with the sun's apparent motion. The principal artists who were employed in the making of clocks to show not only mean but also apparent time, a more curious than useful part of horology, were Sully, Father Alexander, Le Bon, Le Roy, Kriegseissen, Enderlin, L'Admiraud, Passement, Rivar, and Graham. The difference between the hour shown by the sundial and by a correct clock is called the equation of time. In order to obviate the inconvenience produced by this irregularity, astronomers imagine a fictitious sun, coming to the meridian as much before or after the real sun as the equation of time is after or before noon, and as this fictitious sun is supposed to move in a course that is the average, or mean, of all the irregularities in the motion of the real sun, it is called the mean sun, and the time it indicates mean time. An astronomical clock is one contrived to show the apparent daily motions of the sun, moon, and stars, with the times of their rising, setting, and the like. The reader will remember that the motions of the sun and stars round the earth are only apparent, and are produced by the real motion of the earth in rotating upon its axis, which motion is uniform and unvariable.

In the church of St. Mary, Warwick, which has a set of tuneful chimes, is a large wooden clock, fixed to the wall on one side at the west end, showing the day of the month, as well as the time of the day. Upon its smartly painted face are the arms of Queen Anne, and her motto, "Semper Eadem" (Ever the Same); and also the admonition, "Vigilate et Orate" (Watch and Pray).

In the South Kensington Museum is an upright clock, the case of which is richly inlaid with marqueterie scroll-work. The dial is marked "Henry Poisson, London." The height of the clock is eight feet three inches and a half. It is of the period of Queen Anne; and it was purchased for 12*l.* In the same museum is a clock in the form of a globe, supported by three bronze amorini. It is of French work, about the year 1700. The height is three feet eight inches and three-quarters; and the diameter, two feet. It was purchased for 150*l.* Mr. G. H. Morland has a clock mounted on a truncated column of ormolu, with a globe and mathematical instruments. It is of French manufacture, and of the eighteenth century. It has been exhibited in the above-named museum. In that national collection is also a clock, the case of which is in old Florentine porcelain, surmounted by a figure of Time. The dial is surrounded by a wreath of olive leaves in ormolu. The height is two feet six inches. This clock is of the eighteenth century, and was purchased for 18*l.* Also an old French clock, in an ormolu case. It is of the period of Louis XIV. The dial is inscribed "Thuret, à Paris." The height is twenty-two inches. The clock was purchased for 15*l.*

The clock-house at Rochester, which occupies the site of the ancient Guildhall, was built at the sole charge of Sir Cloudesley Shovel, in the year 1706. The clock was also the gift of that distinguished Admiral.

The will of Bishop Burnet, dated October 24th, 1711, contains the following items of clocks and watches belonging to him:—"The clock in the parlour at Salisbury;" "my repeating-watch;" "the clock in the room before my study in Salisbury;" "the repeating-table-clock tipt with silver;" and "the clock in the parlour at St. John's."

Again we have occasion to introduce a bit of morality upon the subject of horology. Prior, who was born in 1644, and died in 1721, thus wrote:—

> " So, if unprejudic'd you scan
> The goings of this clockwork man;
> You find a hundred movements made
> By fine devices in his head;
> But 'tis the stomach's solid stroke
> That tells this being what's o'clock."

A correspondent of 'Notes and Queries,' of September 16th, 1865, says:—" Some four or five years since, on entering a loft in a coal-wharf in this town (Bodmin), my attention was drawn to an antique clock silently standing on a bracket, and begrimed with dust and dirt. It was without a case; the pendulum and weight uncovered like a Dutch clock; the bell formed a dome above. It had the inscription 'William Allmand in Loutheberry fesitt.' The grimy tenant of the loft told me that it was the property of his employer, and that it went by the name of 'Bishop Hall's clock.' On account of its ancient look, I bought it of the owner, and received with it the following history. It was formerly in the possession of the Rev. Robt. Walker, of South Winnow, in Cornwall, and was valued by its owner as 'Bishop Hall's clock.' After Mr. Walker's death, his household goods were sold, and this clock was then purchased by the coal-merchant. I subsequently found that this Mr. Walker was a descendant of Hall, the famous Bishop of Exeter, and afterwards of Norwich."

On the west side of St. John's Lane, Clerkenwell, is situated Albion Place, which was erected in 1822, on the site of an old court, called St. George's Court, which was then pulled down. Here in 1721 lived Christopher Pinchbeck, the discoverer of an ingenious alloy of metals, closely resembling gold, which was named after him, "Pinchbeck," and the inventor of "astronomico-musical clocks." He appears to have excelled in the construction of musical automata, which on several occasions he exhibited in a booth at Bartholomew Fair; and in conjunction with Fawkes, the conjuror, at Southwark Fair. He made a musical clock for Louis XIV. of France, which is said to have been an exquisite piece of workmanship, and worth about 1,500*l*. He also made a fine organ for the Great Mogul, worth 300*l*. He died on November 18th, 1732. From the following advertisement, which appeared in 'Applebee's Weekly Journal,' of July 18th, 1721, it appears that Pinchbeck removed from Clerkenwell to Fleet Street about that time:—

"Notice is hereby given to Noblemen, Gentlemen, and Others, that Chr. Pinchbeck, Inventor and Maker of the famous Astronomico-Musical Clocks, is removed from St. George's Court, St. Jones's Lane, to the sign of the 'Astronomico-Musical Clock' in Fleet Street, near the 'Leg' Tavern. He maketh and selleth Watches of all sorts and Clocks, as well for the exact Indication of Time only, as Astronomical, for shewing the various Motions and Phenomena of planets and fixed stars solving at sight several Astronomical problems, besides all this a variety of Musical performances, and that to the greatest Nicety of Time and Tune with the usual graces; together with a wonderful imitation of several songs and Voices of an Aviary of Birds so natural that any who saw not the Instrument would be persuaded that it were in Reality what it only represents. He makes Musical Automata or Instruments of themselves to play exceeding well on the Flute, Flaggelet, or Organ, Setts of Country dances, Minuets, Jiggs, and the Opera Tunes, or the most perfect imitation of the Aviary of Birds above mentioned, fit for the Diversion of those in places where a Musician is not at Hand. He makes also Organs performing of themselves Psalm Tunes with two, three, or more Voluntaries, very Convenient for Churches in remote Country Places where Organists cannot be had, or have sufficient Encouragement. And finally he mends Watches and Clocks in such sort that they will perform to an Exactness which possibly thro' a defect in finishing or other Accidents they formerly could not."

The following note was made by George Vertue, the celebrated engraver:—"On Thursday evening, Oct. 4, 1722, being in company, and some talking of curiosities in art, mentioned a fine and curious clock made by Pinchbeck, which, with a small movement or touch, could play many and various sorts of tunes, imitating many sorts of instruments, several birds, &c., the music being just, regular, and tuneable, and the time well observed."

A mezzotinto portrait of Christopher Pinchbeck, by Faber, from a painting by Wood, represents him with an open watch in his hand. His portrait was also published in his shop-bill, oval, folio. His son appears to have carried on his father's business; for, in the 'Daily Post' of July 9th, 1733, appeared the following advertisement:—"To prevent for the future the gross Imposition that is daily put upon the Publick, by a great Number of Shopkeepers, Hawkers, and Pedlars, in and about this Town, Notice is hereby given, That the ingenious Mr. Edward Pinchbeck, at the Musical Clock in Fleet-street, does not dispose of one Grain of his curious

Metal, which so nearly resembles Gold in Colour, Smell, and Ductility, to any Person whatsoever; nor are the Toys made of the said Metal sold by any one Person in England except himself; therefore Gentlemen are desired to beware of Imposters, who frequent Coffee Houses, and expose to Sale Toys pretended to be made of this Metal, which is a most notorious Imposition upon the Publick. And Gentlemen and Ladies may be accommodated by the said Mr. Pinchbeck with the following curious Toys, *viz.* Sword Hilts, Hangers, Cane Heads, Whip Handles for Hunting, Spurs, Equipages, Watch Chains, Coat Buttons, Shirt Buttons, Knives and Forks, Spoons, Salvers, Tweezers for Men and Women, Snuff Boxes, Buckles for Ladies' Breasts, Stock Buckles, Shoe Buckles, Knee Buckles, Girdle Buckles, Stock Clasps, Necklaces, Corrals. And in particular Watches, plain and chased, in so curious a Manner, as not to be distinguished by the nicest Eye from real Gold, and which are highly necessary for Gentlemen and Ladies when they travel; with several other fine Pieces of Workmanship of any Sort, made by the best Hands. The said Mr. Pinchbeck likewise makes Astronomical and Musical Clocks; which new invented Machines are so artfully contrived as to perform on several Instruments great variety of fine Pieces of Musick, composed by the most celebrated Masters, with that Exactitude, and in so beautiful a Manner, that scarce any Hand can equal them. They likewise imitate the sweet Harmony of Birds to so great a Perfection as not to be distinguish'd from Nature itself. He also makes Repeating, and all other Sorts of Clocks and Watches; particularly Watches of a new Invention, the Mechanism of which is so simple, and the Proportion so just, that they come nearer Truth than any others yet made. He also mends all Sorts of Clocks and Watches, Musical Machines, and Pieces of Machinery whatsoever, after so just a Manner, that they shall go well tho' they never did before." At the top of this advertisement is a rude woodcut, representing a table-clock.

In the Historical Chronicle of the 'Gentleman's Magazine' of June, 1765, we read, that on May 4th, in that year, Messrs. Pinchbeck and Norton set up at the Queen's House a new clock with four faces, which was greatly admired. The first and principal face showed the true and apparent time, with the rising and setting of the sun every day in several parts of the world, by a moving horizon; the second front showed the motions of the planets in their orbits, according to the system of Copernicus; the third, the age and different phases of the moon, with the time of the tide at thirty-two different seaports; and the fourth, by a curious retrograde motion in a spiral, showed every day of the month and year, with the months and days of the week in proper emblems.

The steeple of Cripplegate Church contains a clock, which presents one face to the east and another to the north. It was made by Lang Bradley in the year 1722, and was thoroughly repaired in 1797 by William Dorrell, of Bridgewater Square, and made to strike the hour on the tenor bell. In 1846 it was again repaired by Thwaites and Reid, and the pendulum was lengthened. This clock is now considered to be one of the best in London. A former alderman, Sir William Staines, among his other munificent gifts to the parish, presented a small and inharmonious bell, upon which the clock strikes a note, discordant to the peal. The chimes, which are almost without a rival, were constructed in 1792 by George Harman, of High Wycomb,

Bucks, and play seven tunes, changing the tune daily at noon. They were thoroughly repaired in 1849, when the tunes were rearranged, and those of inferior melodies taken off, and others substituted.

Clocks have been applied to other purposes than the mere simple measurement of time. George Graham, an eminent horologist, of whom further mention will be made presently, applied the movement of a clock, showing sidereal time, to make a telescope point in the direction of any particular star, even when below the horizon. In Greenwich Observatory was set up a curious transit clock, made by Graham, but greatly improved by Earnshaw, who so simplified the train as to exclude two or three wheels, and also added cross-braces to the gridiron pendulum, by which an error of a second per day, arising from its sudden starts, was corrected. Sidereal time, or star time, is that which is obtained by the observation of the transit of stars over the meridian, and, as it depends upon the rotary motion of the earth alone, and is uninfluenced by the irregularities of orbital motion, it is subject to no variation, and is consequently most uniform. But there is a considerable difference between solar and sidereal time, the sidereal day being four minutes shorter than the solar day.

In the 'Philosophical Transactions' of 1726 is an account of a contrivance to avoid the irregularities in a clock's motion occasioned by the actions of heat and cold on the pendulum rod, by George Graham. And here we will explain that a pendulum in order to maintain at all times an invariable rate of vibration must always remain of an invariable length; but because the changing temperature of the atmosphere alternately lengthened and shortened the pendulum, by the effects of expansion and contraction on the materials of which it was composed, it became necessary so to construct it that while the expansion of one part would make it longer, the simultaneous expansion of another part would make it shorter, and the two antagonistic expansions thus compensating each other, the same length would be maintained at all temperatures, the effects of contraction being corrected in the same manner. A number of methods, some of which Graham invented, have been used, with more or less success, for effecting this object. Graham was an ingenious mechanic, and a most accurate clock and watch maker. He was born in Cumberland in 1675, and died in 1751. The following obituary notice of him appeared in the 'Daily Advertiser' of November 18th, 1751:—"Saturday Evening, died suddenly, at his House in Fleet Street, Mr. George Graham, not less known in the learned World, than in that Branch of Business to which for many Years he had so successfully applied himself, as by his uncommon Ingenuity to have acquired the Reputation of being the best Watchmaker in Europe. He was many Years Fellow and one of the Council of the Royal Society. His Apparatus, made for measuring a Degree of the Meridian in the Polar Circle, is greatly esteemed among the Literati; as are also his many curious Instruments for Astronomical Observations. He lived beloved, and died universally lamented." Graham's portrait, engraved by Faber in mezzotinto after Hudson, represents him sitting with his hat on his knees. Another portrait engraved by Ryley was published in 1820, by Laurie, of No. 53, Fleet Street, after an original picture in the possession of the Earl of Macclesfield.

In 1727 William Hogarth published a satirical print called the 'Masquerade Ticket,' on the top of which was a clock, that is thus described in John Ireland's 'Hogarth Illustrated,' in which it is represented (*vide* also the 'Mirror,' vol. xi., p. 377): "The head of the renowned Heidegger, master of the mysteries and manager in chief, is placed on the front of a large dial, fixed, lozenge fashion, at the top of the print, and, I believe, intended to vibrate with the pendulum; the ball of which hangs beneath, and is labelled Nonsense. On the minute finger is written Impertinence, and on the hour hand Wit; which seems to intimate, nonsense every second, impertinence every minute, and wit only once an hour! The time is half-past one—the witching hour of night. Seventeen hundred and twenty-seven, the date of the year this print was published, is on the corners of the clock. Recumbent on the upper line of the print, and resting against the sides of the dial, the artist has placed our British lion and unicorn renversé (such, I think, is the term in Heraldry), lying on their backs, and each of them playing with his own tail. The lion sinister, and the unicorn dexter, the supporters of our regal arms, being thus ludicrously introduced, may, perhaps, allude to the encouragement King George the Second gave to Heidegger, who at that period might be said to 'teach kings to fiddle, and make senates dance,' who, by thus kindly superintending the pleasures of our nobles, gained an income of 5,000*l.* a year, and, as he frequently boasted, laid out the whole in this country. Under the clock, Hogarth delineated the scene of a masquerade."

In 1730 appeared the following advertisement of a clock-lamp: "Walker's new invented Clock Lamp, which not only far exceeds any Lamp hitherto invented both as to its Neatness in the using (it neither daubing the Fingers at all in the Dressing, nor making any

Daub where it stands, as others do) as well as in the clear continued Light it gives from the first lighting of it till all the Oil is quite spent, without so much as once snuffing it the whole Time; but it likewise shews the Hours of the Night exactly as they pass, from the Time of going to Bed till the next Morning; or, from the Time of lighting the Lamp till it is quite burnt out, supplying, at once, the Place of a Clock or Watch and Candle, when used with proper Oil, which is sold with the Lamps, wherein it is entirely, and beyond all others; and, being conveniently placed by the Bedside, one may lie still in Bed and see how the Time passes. Sold only by John Walker, removed from Fleet Street to the White Horse and Bell, near Cheapside Conduit; where is also sold all sorts of Brazier's and Ironmonger's Goods and the newest fashion'd French Plate very cheap, he being the Maker." This artist appears to have had a formidable rival in trade; for we find the following advertisement appearing in 1731: "Ashburne's New Invented Clock-lamp, shewing the Hours of the Night, exactly as they pass, far exceeding anything of this Kind ever yet invented. To be sold Wholesale or Retale by the Inventor and Maker, Leonard Ashburne, at the Sugar Loaf in Pater Noster Row, next Cheapside; at Mr. Cole's, at the Rainbow and Dove in Chancery Lane, next Fleet Street; at Mr. Noon's, at the White Hart in the Poultry; at Garraway's Old Shop, the South entrance of the Royal Exchange; and at Mr. Taylor's, a China Shop, over against St. Alban's Street in Pall Mall. Where likewise are Sold, All Sort of Chamber Lamps, in the greatest Variety, which have given the Publick a general Satisfaction. With Oil fit for the Purpose. Whereas a malicious and ill designing Advertisement was set forth by the Lamp Pretender in

Cheapside in the Craftsman of Saturday last, and in the Daily Post of Wednesday the 6th Inst. wherein he calls those Lamps sold in Pater Noster Row counterfeit, &c., which any Person on View will find to be false and malicious, endeavouring thereby to lessen the Character and Reputation which they have gain'd beyond all others, notwithstanding his imposing on the Publick, as being the Maker, and now within this Week stiles himself the Inventor, when he knows, as do many others, that they were made long before he was concerned in them. And as for counterfeiting them, I desire that all persons that shall have any Occasion, will do themselves the Justice to see both sorts, before they will be imposed on by such a base and ill designing person." At the top of this advertisement is a rude woodcut, representing four lamps, one of which is somewhat like a coffee-pot in shape; the others are formed like vases; and in all, the part which holds the wick projects from the side of the lamp.

In the vestry books of Bampton, Oxfordshire, is the following entry, under date January 27th, 1733:—" At a vestry this day held, and application being made to the said vestry by John Reynolds, of Hagbourn in the county of Berks, blacksmith, for payment of the sum of thirty-four pounds, due to him for making a new clock and chimes in the parish church of Bampton, he having performed his said work according to his agreement, and to the satisfaction of this vestry, therefore it is ordered by this vestry, that the churchwardens of this parish for the time being do forthwith pay unto the said John Reynolds the said sum of 34*l.*, according to agreement of this vestry for that purpose, except 40 shillings, which is to be left as a caution till the clock is further proved." Bampton Church Tower now contains a large clock, which, when it strikes, is heard to a considerable distance. There are also chimes, which play an ancient carol, at the hours of one, five, and nine.

During the eighteenth century musical and automaton clocks were much in fashion. In addition to those which we have before referred to, we may mention the following. In Bagford's Collections, Harl. MSS., 5946, in the British Museum, is this advertisement:—"The celebrated musical clock, approved of by the greatest quality in this kingdom, which, besides divers curious motions, performs, 1. A Consort of Italian and English music, either single, or in parts, to the number of 32 different tunes, including sets of airs, minuets, jiggs, borees, sarabands, courants, &c., on organs, trumpets, flutes, and flagellets, very true and melodious. It shifts a fresh tune of itself, and repeats at pleasure. 2. In the course of this harmony, the seven liberal sciences, *viz.* Musick, Optick, Physick, Architecture, Painting, Mathematicks, and Eloquence, appear, each with some proper instrument to denote her profession. 3. Apollo breaks through a cloud with his harp in his hand. 4. A cuckoo calls, and 17 small birds warble their proper notes as natural as if living. Its model is exact and regular, and adorned with a vast number of curious carved and gilded figures, with the Queen under a triumphal arch, the pedestal the same as at St. Paul's, and many other curiosities. To be seen at the 'Duke of Marlborough's Head' in Fleet Street, from nine in the morning till eight at night. To be sold for 700 guineas, or raffled for at five guineas per ticket." The King of Spain, before the French revolution had a clock, which played one hundred tunes loudly like a full band. The bottom part was a large stand; above on a pedestal were three ormolu figures, about four feet high; on the top was a large globe representing the earth, which moved so as to display the diurnal and also the annual motion of the world. This clock came into the possession of a Madame Beauzalie, of Paris, who asked a very large price for it. There is a musical clock at Knole, Kent, with a bird at the top of it, which sings and flaps its wings. At the same place is also a fine clock, which was presented by Louis XVI. to Lord Whitworth, the second husband of the Duchess of Dorset. While mentioning Knole we may add that the out-of-doors clock there formerly stood on a dome, similar to that at Lambeth Palace, over the Hall; but the roof of that room appearing to bend under it, it was taken down in 1745, and removed to its present situation.

In 1740 appeared the following advertisement of an automaton which told the hours, &c.:—"This is to give Notice, To all Gentlemen, Ladies, and others. That the three surprising Mathematical Statues, which has lately been shown before the Vice-Chancellor and the rest of the University at Oxford, with universal Applause, &c. The first Statue represents a Country Lass, with a Pidgeon upon her Head, holding a Glass in her Hand, which she lifting up, makes to run out of the Pidgeon's Bill, either Red or White Wine. The second represents a Merchant Grocer in his Shop, who at the Word of Command, opens and shuts his Door himself, and brings Sugar, Tea, Coffee, and all Sorts of Spices. The Third Statue represents a Blackmoor holding a Hammer in his Hand, having a Bell before him, and as soons as commanded performs as follows, *viz.* 1. He strikes the Day of the Week, and the Hours. 2. Four Spectators may each of them draw a Card, and he will not only shew how many spots are upon such a Card, but also the Colour of the Card. 3. Any of the Spectators may call for any Number under 12 in his Mind, and the Blackmoor will strike it. With many other Actions too tedious to mention. To be seen from Ten in the Morning until Eight in the Evening, at the End of Wine Office Court in Fleet Street. Price 1s."

On July 15th, 1740, appeared the following descriptive advertisement of a musical and astronomical clock:—"Now open'd, the Athenian Temple of Arts and Sciences, consisting of a great Variety of fine Performances in Musick and other Arts, being the Produce of a large Expence and Ten Years Labour and Study, is now finish'd, and is humbly offered to the Ingenious and Polite, and to all true Lovers of Arts and Sciences, of both Sexes, for their Approbation, and is to be seen and heard, 'till Sold, in the North Isle of the Royal Exchange, at 2*s.* 6*d.* each Person. The Description. This Piece of Machinery is in Form of a Roman Temple, made of Mahogany, finely Embellish'd with Emblematical Carve Work gilt, and fine Poetical Paintings. The upper Part is a Dome, supported by 24 Columns, in the Concave of which is a Ball, representing the Sun, with his Six Planets moving round him in proportionable Periods of Time; underneath which is the Statue of Euclid turning round and commanding 12 Emblematical Figures, each performing their several Motions. In the first Front there are four Sets of Keys for the Practice of the Science of Musick, and a curious Piece of Painting over ditto. In the second Front is the Representation of Day by Phœbus, driving the Night before him when he rises. In the third Front are several Scenes of fine moving Paintings, &c., by Clockwork. In the fourth Front is a Scene of Painting, representing Night, which moves off, and you discover a curious Clock that performs a fine Air, or Piece of Musick, at set Hours, or at Pleasure. The Clock is of a new Invention, finely embellish'd, and solves many and curious Problems relating to Astronomy, the Particulars of which are too many to be express'd in this narrow Compass. This Piece of Machinery performs many delightful Pieces of Musick upon the full Organ by Clockwork. The Entertainments of Persons with the Performance of this Grand Temple, will begin at 11, 12, 3, 4, 5, 6 of the Clock every day. Note, Any select Company may be privately entertain'd at any Hour separate from the Time above."

John Whitehurst, of Derby, writing in 1847, says:—"The escapement alluded to in my specification (for the Westminster clock) is now keeping time in a church clock in the neighbourhood. It is the very best for large turret clocks that I know of in all my many years' experience. There is one of the same escapements in All Saints Church clock, Derby, made by my late great-uncle, John Whitehurst, F.R.S. The clock was made by him A. D. 1745, and the clock is now going to time most admirably. I have made a number of large turret-clocks with the same escapement, and they invariably keep excellent time." John Whitehurst, an eminent philosophical and mechanical genius, having settled at Derby as a watchmaker, made the clock and chimes of the beautiful tower of All Saints Church as above mentioned. From his vicinity to the many stupendous phenomena in Derbyshire, which were constantly presented to his observation, he was excited to investigate their causes. Being appointed stamper of the money weights in 1775, he removed to London, where he soon after published his 'Enquiry into the

Original State and Formation of the Earth,' which will remain a monument of his fame to future ages. Particulars of his, and of another watchmaker's, business relations with Wedgwood, in his pottery art, will be found in Miss Meteyard's 'Life of Josiah Wedgwood.' He died in 1788, at his house in Bolt Court, Fleet Street, where Mr. Ferguson, another celebrated self-taught philosopher, had then recently lived and died. His death was universally lamented.

The 'Gentleman's Magazine' for December, 1746, tells us that the following was the inscription on a clock belonging to Mr. H——s, at H—ll—ns, in Yorkshire, in that year:—

> "I serve thee here, with all my might,
> To tell the hours by day, by night:
> Therefore example take by me,
> To serve thy God, as I serve thee."

Similar lines were posted on his house-clock by the Rev. John Berridge, Vicar of Everton, who was born in 1716, and died in 1793, and whose name is familiar to those who know the history of the revival of evangelical religion in England in the last century:—

> "Here my master bids me stand,
> And mark the time with faithful hand;
> What is his will is my delight,—
> To tell the hours by day and night.
> Master, be wise, and learn of me,
> To serve thy God as I serve thee."

The following lines, which are inscribed under the clock in front of the town-hall in Bala, Merionethshire, are in part like the above:—

> "Here I stand both day and night,
> To tell the hours with all my might;
> Do you example take by me,
> And serve thy God as I serve thee."

Over the market clock at Terni is this moral notice: —"Hora, dies, & vita fugit, manet unica Virtus."

"Hours, days, and ages fly away;
Virtue alone knows no decay."

Early in the present century there was in the town of Tetbury, Gloucestershire, a very ancient market-house, in front of which was a clock with a very curious and elaborately carved oaken dial-plate, with this motto:—"Præstant æterna Caducis."

Under or above the dial of the parish clock of Market Harborough, in Leicestershire, appears the excellent injunction:—"Begone about your business."

The 'Gentleman's Magazine' for 1748 tells us of a machine which was suggested by the Marquis of Worcester, and which was to make a ball of any metal, when thrown into water, presently rise from the bottom, and constantly show by the superficies of the liquid the hour of the day or night; never rising more out of the water than just to the minute it showed of each quarter of the hour; and if by force the ball should be kept under water the time was not lost, but recovered as soon as it was permitted to rise to the superficies of the water. This was one of the many projected chimerical inventions of the ingenious Marquis, which he called "Scantlings," meaning probably chips of genius, and which he laid before members of Parliament, from whom, however, he did not receive any encouragement. The same magazine for 1750 records a plan for the construction of a nocturnal horologe, in which the gradual falling of sand through an aperture in a vessel drew up a weight, along a scale marked with notches for the hours and half-hours, which weight and notches being touched in the dark indicated the time.

In 1752 the old clock at York Cathedral, grown useless by age, and whose very large Gothic case covered the wall between the south door and the chapel for early prayers, and blocked up one of the windows, was removed; and an elegant new clock by John Hindley, of York, was put up in its stead. At the same time there was outside the Cathedral, over the south entrance, a handsome dial, both horary and solar; on each side of which was a wooden statue in armour of the time of Henry VII., which struck the quarters on two small bells. Early in the present century these ancient clock and figures were removed, and the present dial was substituted. The clock of St. Martin's Church, Coney Street, York, is illuminated, and projects into the street, and upon it is the figure of a man taking a solar observation.

In 1753 was published a description of two methods by which the irregularities in the motion of a clock, arising from the influences of heat or cold on the rod of the pendulum, might be prevented, with some papers on the same subject previously read before the Royal Society, by John Ellicott, clockmaker and F.R.S. In the 'Philosophical Transactions' of 1761 are 'Observations for proving the going of Mr. Ellicott's clock at St. Helena,' by Charles Mason. In the same work for 1762 appear further observations 'concerning the going of Mr. Ellicott's clock at St. Helena,' by James Short, F.R.S. In the same work and year were published 'Observations on a Clock of Mr. John Shelton, made at St. Helena,' by Nevil Maskelyne, D.D., F.R.S., an astronomer and mathematician. About this time were carried on smart contests on the subject of the improvements made in the pendulum, and several claims to new inventions therein were set up by rival mechanics. In

1772 was published the mezzotinto three-quarter length portrait, engraved by Dunkarton, after Dance, of John Ellicott, who died in that year, aged sixty-seven. He is represented sitting. At Greenwich Observatory was an astronomical clock made by Shelton, before named.

The Horse Guards clock was originally made by Thwaites, of Clerkenwell, in 1756. It is a large thirty-hour clock, striking the quarters upon two bells, and showing the time upon two white dials with black figures, seven feet five inches in diameter, one facing St. James's Park, and the other Whitehall. The frame is of wrought iron, the wheels are of the best yellow brass, and the escape and centre-wheel, arbors, and pinions are of case-hardened and tempered steel. The going part discharges the hours as well as the quarters. Originally the clock had a common recoil escapement, and the pallets became much worn. The pendulum was eight feet two inches long, and to reduce the arc of vibration it was furnished with fans. In 1816 the going part, including the dial, works, and hands, and the connecting rod-work, was made new by B. L. Vulliamy, the Queen's clockmaker. The new going train consists of only three wheels, namely, an escape, which makes one revolution in four minutes; a centre-wheel, one revolution in an hour; and a great wheel and two pinions, each of twenty-two teeth. The brass barrels and caps are new, and a ratchet attached to the great wheel keeps the clock going while being wound. The pivots work in gun-metal holes, in bosses, which are held by screws, not riveted. The front pivot of the escape wheel carries a seconds' hand, which, as the pendulum vibrates two seconds, shows the seconds by the hand advancing two seconds each vibration. The pendulum has a teak-wood rod, and a cast-iron bob, weighing one hundred and ninety pounds. It is terminated by a gun-metal screw and nut, cut into a strong thread, by which the clock is regulated. The pendulum is so hung as to be independent of the frame of the clock, which can be taken to pieces and cleaned without disturbing it. The escapement is the pin-wheel. In this clock, says Vulliamy, are applied, it is believed for the first time, the following improvements:—a dead pin, wheel escapement, a two seconds' pendulum, a ratchet, to keep the clock going while being wound, and a degree plate, to indicate very correctly the arc of the vibration of the pendulum.

And here we may mention that the accuracy of a clock depends chiefly on its escapement, or that part of the mechanism which connects the regulating power with the wheelwork. Babbage, in his 'Economy of Manufactures,' 1835, tells us that "clocks and watches may be considered as instruments for registering the number of vibrations performed by a pendulum or a balance. The mechanism by which these numbers are counted is technically called a scapement. It is not easy to describe; but the various contrivances which have been adopted for this purpose are amongst the most interesting and ingenious to which mechanical science has given birth. Working models on an enlarged scale are almost necessary to make their action understood by the unlearned reader; and unfortunately these are not often to be met with. A very fine collection of such models exists amongst the instruments at the University of Prague."

The 'Annual Register' for 1760 tells us, that at a quarterly meeting of the governors of Dublin workhouse, held on October 6th, 1760, under the presidency of the Earl of Lanesborough, it was resolved unanimously, "That the thanks of this board be presented to the Right Hon. Lady Arabella Denny, for the continuance of her kind and most useful attention to the foundling children, particularly for a clock lately put up in the nursery, at her Ladyship's expense, with the following inscription, *viz.* 'For the benefit of infants protected by this hospital, Lady Arabella Denny presents this clock, to mark, that as children reared by the spoon must have but a small quantity of food at a time, it must be offered frequently; for which purpose, this clock strikes every 20 minutes, at which notice all the infants that are not asleep must be discreetly fed.' "

Repeating clocks and watches are instruments for registering time, which communicate their information audibly only upon the pulling of a string, or by some similar application. Several instruments have been contrived for awakening the attention of the observer at times previously fixed upon. The various kinds of alarums connected with clocks and watches are of this kind. In some instances it is desirable to be able to set them so as to give notice at many successive and distant periods of time, such as those of the arrival of certain stars on the meridian. A clock of this kind was used at the Greenwich Observatory a long time ago. Ferdinand Berthoud, a French marine clockmaker, who was born at Plancemont, in Neufchâtel in 1727, and died in 1807, invented an astronomical clock, which by means of a cord was made to strike the seconds during the progress of an observation; the number of beats being counted gave the time of duration. Berthoud wrote several works on horology in his own language, some of which were reprinted. The first appeared in 1760.

The subjoined description of a curious vocal clock is

given in the journal of the Rev. J. Wesley:—"On Monday, April 27, 1762, being at Lurgan, in Ireland, I embraced the opportunity which I had long desired, of talking to Mr. Miller, the contriver of that statue which was in Lurgan when I was there before. It was the figure of an old man standing in a case, with a curtain drawn before him, over against a clock, which stood on the opposite side of the room. Every time the clock struck, he opened the door with one hand, drew back the curtain with the other, turned his head as if looking round on the company, and then said, with a clear, loud, articulate voice, past one, or two, or three, and so on. But so many came to see this (the like of which all allowed was not to be seen in Europe), that Mr. Miller was in danger of being ruined, not having time to attend to his own business. So as none offered to purchase it or reward him for his pains, he took the whole machine to pieces."

About the middle of the eighteenth century, Colonel Magniac, a famous clockmaker, lived in an old mansion in St. John's Square, Clerkenwell, and had his workshops there. This manufacturer's automaton clocks caused much interest at the Court of China and elsewhere, and perhaps did much to render Clerkenwell noted as the clockmaking parish. Two of the most remarkable clocks manufactured by him for the emperor were rare specimens of mechanical skill; in addition to regiments of soldiers, musical performers parading, beasts and birds, all displaying appropriate action, combined to show what varied and graceful motions could be produced by wheels, pinions, and levers, and while pleasing the eye also charmed the ear by the bell music, tunes, and chimes. Early in the present century the above-mentioned mansion was pulled down, and Messrs.

J. Smith and Sons purchased the freehold and built their extensive clock manufactory upon it. The workshops and warehouses of this establishment are the largest of any used for clockmaking in Clerkenwell.

The 'Annual Register' for 1764 informs us that som time before that year the Royal Academy of Paris was presented with a repeating clock, which struck the hours and quarters with only a single striking wheel, the inventor rejecting two-thirds of the pieces contained in the striking part of the ordinary repetition clock.

In the last century, Henry Bridges, an architect, of Waltham Abbey, constructed an elaborate mechanism called the Microcosm, or Little World; which, after his death, came into the possession of one Edward Davis, who, some time between 1764 and 1789, publicly exhibited it, and published a pamphlet describing it, containing an engraving of the machine. It was built in the form of a Roman temple, being ten feet high, and six feet broad at the base; and consisted of five principal parts. It contained a variety of moving, life-like figures of men, women, and animals. At the top were three scenes, which changed alternately. The first represented the Muses on Parnassus; the second, a forest with Orpheus and wild beasts; the third, a sylvan grove with birds flying and singing.

In the centre of the structure stood a clock, which was divided into two astronomical systems, the uppermost being the Ptolemaic, and the lower the Copernican. In the former, the earth represented the centre of motion, with an hour circle round it, and the sun and moon revolved every twenty-four hours. It showed the rising, southing, and setting of the sun, which always pointed out the hour of the day, and rose and set at the same time as it really did. There were two blue circular plates, called horizons, one on the left hand, and the other on the right, which rose and fell according to the lengthening or shortening of the days, in order to regulate the sun's rising and setting; so that as the sun went round in twenty-four hours, being within the hour circle, it passed by every hour on the plate, and pointed out the time of the day. In order to tell the time in the night after the sun was set, there was an hour-hand placed directly opposite to the sun, also a minute and second hand, all on one centre, so that the time of the day or night might be as easily known by this as by any other clock. There were two other hands on this centre, one called the moon's nodes, which showed the time of the eclipses; and the other the equation of time, or the difference between the sun and the clocks. There were also two lesser horizons within those of the sun, to regulate the rising and setting of the moon, in the same manner as those of the sun. The moon showed the time of new and full, her increase and decrease, always representing the same figure as it really was in the heavens.

The Copernican system was composed of seventeen bodies, in the centre of which was placed the sun, showing its rotation, and the planets moving round it, in the same signs of the zodiac as they were in reality. Their orbits had all due inclinations, and lay in their true plains. Underneath were folding doors, which on being opened discovered four other systems, in which the motions of the planets were more clearly described. The first was the solar system; the second, the system of Jupiter and his moons; the third, the eclipses of the sun and moon; and the fourth, the earth's annual motion. Another piece of mechanism represented the course of the comet of 1759.

Below these was the representation of a landscape and the sea. On the former vehicles, persons, and animals were seen in motion; and on the latter ships were sailing. Another scene displayed a carpenter's yard with men at work. During the whole performance of the machine, when upwards of one thousand two hundred wheels and pinions moved all together, it played several pieces of music on an organ, which likewise was provided with keys, so that any piece might be played upon it.

Again we introduce a passage from the poetry of horology. Young, who was born in 1684, and died in 1765, wrote these suggestive lines:—

> "From old Eternity's mysterious orb
> Was Time cut off and cast beneath the skies—
> The skies which watch him in his new abode,
> Measuring his motion by revolving spheres,
> That horologe machinery divine."

Stollenwerk Clock at Stowe — Shakespeare Clock — Renaissance Clock — Wooden Clock made by a Belgian Peasant — Clock Time Map — Old Clocks and Old Clockmakers — Clocks in the Great Exhibition of 1851 — Alpha Clock — Clock Bed — Shepherd's Electric Clock — Clock from Paris Exhibition of 1855 — Westminster Palace Clock — Clock Omen — Huge Clock Bells — Musical Clock Bed — Perpetual Clock — Clock Lock — Self-Winding Clock — Camel Clock from Pekin — Clocks in the International Exhibition of 1862 — Clocks without Dials — Dutch Clocks — Dutch Clockmakers — Clock Chimes in the Low Countries — Antwerp Cathedral Chimes — Belgium Clocks — American Clocks — The Old Clock on the Stairs — French Clocks — Japanese Clocks — Atmospheric Clock — Clock without visible Works — Clocks imported into England — Proverbs relating to Clocks — Folk Lore of Clocks — The Legend of Blownorton Clock — Tempus Fugit.

At the Stowe sale in 1848 a magnificent Huygens clock, made by Stollenwerk, was sold to Mr. Paxton for only fifty-one guineas, although it was said to have cost the Duke of Buckingham one thousand. The case was of the most exquisite marqueterie, and was also enriched with well-executed figures of boys, trophies, and masks in ormolu. The outline of this choice article was thoroughly good, and the beauty of the several decorations renders the small sum that it fetched remarkable.

At an exhibition of select specimens of British manufactures and decorative art, given by the Society of Arts in 1848, Summerly exhibited the Shakspeare Clock, which was designed and modelled by Bell, and made in Parian by the Mintons; the works were to be furnished by Vulliamy; the dial was to have Drayton's silvering process; and was placed between two figures representing Tragedy and Comedy, as typical of Time passing between Joy and Grief:—

> "Joy absent, grief is present for that time."—*Richard II.*

The composition was surmounted by a statuette of Shakspeare, the likeness being founded upon his bust in the chancel of Stratford Church.

At an exhibition of ancient and mediæval art, given by the Society of Arts in 1850, Mr. Baring Wall, M.P., exhibited a curious little silver-gilt clock in the style of the Renaissance. It was surmounted by a figure of Cupid, who held a very long arrow, with which he pointed to the hours marked upon the edge of a dome on which he was seated. The dial circle was at the top of the clock, below the Cupid, and not at the side as in ordinary clocks.

In a number of the 'Brussels Herald' for the year 1851, appeared the following account of an ingeniously constructed wooden clock, which was made by a poor Belgian peasant, and by him called "The Moving Calendar:"—"There are several dials to mark the hours, minutes, seconds, and days of the week, the day of every month, the months of the year, the years and centuries, the rising and setting of the sun, moon, &c. The works are encased in glass, and are so arranged that they may be inspected, leaving no doubt on the mind as to the execution of the workmanship or regularity of the movements. About the dials and galleries there is a gallery about a yard in length, with cells in the centre, and a tower at each end. When the clock is about to strike, the door of one of the cells opens, and the figure of Death appears armed with a scythe, followed by the figure of our Saviour with a whip in his hand, who drives the enemy before him, and finally thrusts him into another cell. At the first stroke of the clock a little cock, perched on the cross of the steeple, flaps his wings and stretches out his neck as if in the act of crowing. As soon as the clock has finished striking, the different figures return to their respective cells, taking care to close the door behind them. Thrice a-day, at six a.m., twelve at noon, and six p.m., by means of ingenious mechanism, the sound of the Angelus is heard. The Holy Virgin leaves her cell, appears on the gallery, and withdraws to an oratory. At the same time an angel descends from one of the towers and places himself under the Virgin, bending in the act of salutation. The Virgin Mary is made to appear disconcerted, trembles, and shows signs of deep emotion, after which the angel resumes his former position and performs the same movement twice. All the works of this admirable machine are of wood or brass, and remarkably well executed. The inventor designed and carried out his plan without assistance. During the day he worked in the fields, and at night, by lamplight in a corner of his garret, he constructed this extraordinary clock."

In 1851 Ellis and Son, watchmakers, of Exeter, published a very interesting map showing the time kept by public clocks in various towns in Great Britain. Among many other curious notes which may be made on this subject, we may mention that it is Sunday in Inverness and Glasgow nearly seventeen minutes earlier than it is at Plymouth; and it will be 1867 in Liverpool eleven minutes before it will be so in Bristol.

A correspondent of 'Notes and Queries' for 1851 said that he had in his possession an apparently very old, although very elegant and excellent, eight-day clock, with the maker's name on its face, Thomas Lestourgeon, London. It was then a century old. Some years previously there was found among the apparatus of the nautical philosophy class in the University of Edinburgh, an instrument called in the inventory "An old watch, maker's name Lestourgeon, London." Another correspondent of 'Notes and Queries' for 1854, stated that he had in his possession, and in use, a clock bearing on its face the name of Orpheus Sumart, of Clerkenwell. The works were of wood, and its mechanism extremely simple. It was more than a century old, and was regarded as a piece of ancient family furniture. The following clockmakers' names are to be seen on various elaborate and curious antique clocks:—Samuel Dunkerley, James Boyce, Aclander Dobson, and John Hallifax, all of London. The latter name is on a very curious chime-clock.

The Universal Exhibition of 1851 was the means of bringing under the notice of the public many rare, beautiful, and unique specimens of the horologist's art, which otherwise might have remained unknown. We shall not attempt to catalogue all the works of this kind which were exhibited; but we will shortly describe some of the most notable among them. The following notes refer to clocks only; we shall describe the watches in due order:—

John Page, of Bury St. Edmund's, exhibited a fine skeleton clock twenty-eight inches in height, which required only once winding in three years. The power was obtained by six springs of the united force of two hundred and fifty pounds, which were enclosed within as many brass barrels; three of the springs were connected by chains to a fusee on the right hand, and three on the left hand. The power was thus concentrated in the pinion of the second great wheel, and was conveyed up the train of wheels and pinions to the pallets which were connected with the pendulum, the power being reduced to that of a few grains; yet it was sufficient to keep the pendulum in constant motion. After the clock was first wound up, the chain was unwound from the smallest part of the fusee, three turns of which allowed the barrel to revolve once in two hundred and ten days. This was believed to be the slowest motion ever produced. By this clock the day of the month and the number of weeks and years during the time it had been going were registered. A spirit-level was attached to the base, so that the clock might be placed for a certainty in a vertical position. Page also exhibited a pyramidical-formed skeleton timepiece, twenty-eight inches high, which went three months with only once winding. The hour-dial was placed at the bottom, in order to display the wheels in motion. It had Graham's dead-beat escapement. The hands were moved by novel but simple mechanism.

William Turner exhibited a clock showing simultaneously the time in London, Edinburgh, Dublin, and Paris. And similar inventions were exhibited by foreign competitors. J. Driver, of Wakefield, exhibited a handsome skeleton clock, which struck on eight bells, and was supported by four lions made of terra-cotta. By this clock the time in any part of the world might be ascertained at pleasure.

A large transparent skeleton spring timepiece was contributed by J. Edwards, of Stourbridge. Some of the wheels were constructed of brass and glass connected together, a novelty in horology.

Sanders Trotman had a curiosity in the shape of a night-lamp time-indicator; the regular consumption of camphine being the power employed to tell by means of an index the time of night.

A portable spring time-keeper, which required only once winding up in four hundred and twenty-six days, was exhibited by J. T. Edwards, of Dudley.

A working shoemaker, resident in London, invented, made, and exhibited a thirty-two day skeleton time-piece, with detached escapement, having fewer works than usual. A tailor exhibited a clock which showed the days and months, the motions of the sun and moon, and the tides in many parts of Europe. It went for twelve months. A blacksmith contributed a musical clock, which played a tune every three hours.

George Taylor, of Wolverhampton, exhibited his perpetual self-correcting date-clocks, in which the day of the month, the month, and the day of the week were all shown in recessed openings under the dial.

J. Briscall, of Birmingham, exhibited a one-month clock. In addition to the ordinary dial, it had an almanack attached, and was self-regulating, so far as the months and the extra day in Leap-year were concerned.

"Tempus rerum imperator" was appropriately painted on the dial of Lovelace's unique astronomical clock, known as the Exeter Clock, which extraordinary example of patient industry and consummate skill on the part of poor Lovelace took him thirty-four years of his life to accomplish. He died in great poverty. This clock was five feet in width, ten feet in height, and weighed about half-a-ton. Two sets of figures in the hour-circle surrounded the dial, namely, from twelve at noon until twelve at night, and from twelve at night until twelve at noon. The shape, size, and age of the moon were very cleverly shown; the month and the day of the month were also indicated. The clock required to be regulated only once in one hundred and thirty years, struck the hours and chimed the quarters. The times of sun-rise and sun-set were also shown by means of an horizon, which receded or advanced according to the season of the year. Once in every four hours a psalm-tune was played, an index at the same time pointing to the name of the tune. During the playing, two moving figures, Fame and Terpsichore, were seen in motion, keeping proper time.

An eight-day striking clock, with only one mainspring, and one train of wheels to do all the work, including the striking, which was performed on one bell placed at the top, was exhibited by William Harvey, of Stirling.

Whishaw exhibited a centrimetal chronometer or velocentimeter, which was intended for ascertaining most accurately merely by observation the velocities of railway engines and trains. A movable ring surrounded the dial, and contained velocities calculated on the standard of one thousand three hundred and twenty feet, or a quarter of a mile. It was furnished with a central dead hundredths hand; so that by moving zero on the ring in a line with the central hand, at the instant of passing one of the quarter-of-a-mile standards and then reading off the marginal figures at the instant of passing the next quarter-of-a-mile standard, the velocity due to the quarter of a mile passed was accurately given. The ring was movable, so that for taking velocities on foreign railways, a kilometral or other scale might be substituted at pleasure.

Roberts's Alpha clock was one of the novelties of the Exhibition. It was called the Alpha from its shape resembling an A. The wheels and pinions were made of cast-iron, with teeth retaining the scale. It had only one weight to actuate both the going and the striking trains; and the chain or cord, requiring no lateral traverse, could be taken off in any direction. The compensation pendulum and the remontoire escapement were adapted to keep the clock at an almost uniform rate: whilst the hands, being advanced at intervals of thirty, or, if desired, sixty seconds, afforded an opportunity for ascertaining the time to a second. The striking of the hour was effected through means by which the blows were given at equal intervals of time, thus avoiding irregularity and expenditure of power to direct it.

Lucien Marchand contributed a curious musical clock, which required winding only once a month. It had independent seconds. On one side was a magician with a wand, who at certain intervals rose from his seat and pointed to an answer to any of twelve set questions which might be proposed, the whole being arranged in connection with the clockwork. On the other side was a bird perched on a tree, which sang at certain intervals. The music was from four popular overtures.

Another artist displayed a timepiece of clockwork consisting of a pillar, the spring capital of which opened when touched, and up jumped a pea-sized bird, which fluttered its wings and chirruped for two minutes.

Frodsham exhibited a very elegant clock, upon a slab supported by an eagle. Above the clock was an ornament composed of a broken pillar, the work of time; and a serpent, the emblem of eternity.

Miss Walter exhibited a curious and fanciful design for a clock-face, in which the virtues and vices were arranged in two distinct circles. The former were the inner, next the eye and ear of Omniscience; the latter were the outer and darker circle, denominated Satan's Kingdom.

The inventor exhibited a clock or alarum bed, which turned its occupant out at any period that he might have previously set the clock and its machinery to. It was represented to be valuable to military men, sportsmen, travellers, and others, and all whose pleasure or duty required early and punctual rising. It was perfectly noiseless, as the alarum did not strike, but gently tilted the sleeper into an upright position.

R. and J. Moore exhibited a large and showy clock, which was elaborate in internal workmanship and outside decoration. It went one month, chimed the quarters on eight bells, struck the hours on a deep-toned bell, and played twelve tunes, shifting by the action of the clock to a fresh tune every hour. The frames of the dial and the steel plate upon which the clock stood were enamelled upon a new principle. The base was carved in walnut-wood. The ornamentation was of mixed character, and rich and effective.

Elkington and Company exhibited the Hours Clock-case, so called from the fact of the face being embellished with a bas-relief representing the twelve hours circling round the clock, which itself had an enamelled dial representing the sun, its centre being a flying phœnix, which we are told was born anew every five hundred years. At the base were two figures, respectively illustrative of repose at evening, and the wakening to labour in the morning. The apex was crowned with a figure of Psyche, or the soul, looking upward, emblematic of eternity. The whole was prettily conceived and pleasingly designed. This work was executed in electro-bronze.

A turret-clock with many improvements, manufactured and exhibited by Smith and Sons, was one of the most striking features of the horological section. The same firm exhibited a four hundred-day timepiece, a detector clock or watchman's timepiece, and clocks for China and Turkey. Many of the clocks exhibited in the foreign department, in addition to the hour, minute, and seconds hands, had also the means of indicating the days of the week and month. In Messrs. Smith's four hundred-day clock, which was worked by a spring and had a mercurial pendulum, were the same useful additions. The detector clock could be made a complete check to irregularity and neglect of duty on the part of watchmen, as their absence could be correctly registered on the dial. The registering apparatus consisted of a revolving circular frame, fitted with springs and steel pins; and its general appearance was that of an ordinary-sized bracket-clock. In the ordinary telltale-clock are a number of pins sticking up round the dial, one for every quarter of an hour; and it is the duty of the watchman on the premises where such a clock is kept to go to it every quarter of an hour, and push in the proper pin, to show his employers the next morning that he has not been negligent of his duty. Each pin admits of being pushed in during a few minutes only, and if not heeded, will be found sticking out, showing the exact time when the watchman was away from his post. In a telltale-clock in one of the lobbies of the House of Commons, the face and pins are enclosed behind a glass; and outside the clock-case is a handle, communicating with a small lever, standing over a part of the circle in which the pins move; and as the pins are carried round in a sort of movable dial, the effect of pulling the handle is to push the pin which comes under the lever every quarter of an hour. In these clocks the pins are made to pass over an inclined plane some hours after they have been pushed in, and in this way are pushed out again. In the clocks for Turkey and China, exhibited by Smith and Sons, it had been found necessary to substitute brass chains for the ordinary gut lines or steel chains, so as to prevent the injurious effects of the climate of those lands. The figures on the dials were those of the countries for which they were constructed. In addition to the above, Messrs. Smith also exhibited the "Uniformity of Time Clock and Telegraph," which was invented by Francis Whishaw, a civil engineer, one of the uses of it being to regulate time between distant places to the hundreth part of a minute, by means of sounds transmitted by electrical agency. It also formed a telegraph, as there were four distinct alphabets and numerous signs and signals distinctly marked in red and black on the annular movable plate which surrounded the dial. There were four hands, which rotated together; one of these was distinguished from the others by being of a light colour, and was called the index hand, as by it the class of signals to be used was indicated. The other hands were used for pointing to the signals, which were thus more quickly given than if only one hand had been used. By two electrical bells, of dissimilar sound, the particular quarter of the dial on which the signals were to be read off was readily understood. Besides the telegraph dial and regulator, there was a second face with the ordinary hands, so that one side might be in the telegraph room of the railway station, while the other faced the booking office.

Shepherd's electric clock, which struck the hours with peculiar solemnity, was a remarkable feature of the Exhibition. In adapting this clock to the external design of the building, Owen Jones ingeniously contrived a plan by which the conventional form of a circle for the face of the clock was done away with, in order that the elevation of the south-end of the transept might not be disfigured. The clock face was a semicircle, having as usual twelve divisions, and the figure 12 was also as usual at the top of the circle, the numbers of 1, etcetera, likewise followed in the usual order; but, as with one hand only the semicircular dial would be left by the hour-hand for intervals of each alternate twelve hours, a second number 6 was added on the west side of the dial, and also a second hour-hand, which pointed to the number 6 on the west side as the first hour-hand left the number 6 on the east side. The hour-circle was twenty-four feet in diameter. The hands were of copper gilt. The minute-hand was sixteen feet long, purposely shortened so as not to descend below the fanlight frame. The thirteen figure-plates, which were of zinc, were secured to, and corresponded in shape with, the intersectional spaces formed by the second semicircular bar from the centre, and the radial bars of the great southern fanlight of the transept. The figures were painted white on a blue ground, in order to harmonize with the two prevailing colours of the external decoration of the building. The effect of Shepherd's improvements in the application of electricity to horological purposes was to attain a greater uniformity and certainty in the going of his clocks. The leading features in the Exhibition clock were the application of electricity to the winding-up of the impulse spring or weight, in order to render the escapement or impulse given certain in its action; and to improvements in effecting the movement of the train, in order to denote the hours, minutes, and other subdivisions of time. In this clock certain alterations in the details of the magnetic apparatus was rendered necessary in order to suit the particular case. Besides the great clock for the transept, two dials of smaller size, one at the east and the other at the west end of the building, were also at work in connection with it. The electrical current to each of these auxiliaries was transmitted through copper wire coated with gutta percha. The mechanism of the clock was fixed in the south gallery of the transept, at about forty-eight feet below the centre of the dial, and motion was communicated to the hands by means of a rod made up of several lengths of brass tubing screwed together, and of one-and-a-half-inch in diameter. The clock frame was much lighter than usual, as the ordinary heavy weights were entirely dispensed with. There were two wheels within the frame, placed vertically, namely, the escape-wheel of ten inches diameter, to which the power was applied, and a larger or central vertical wheel, of eighteen inches diameter, working into the pinion on the arbor of the escape-wheel, which was in two parts, the teeth of each part being placed in opposite directions; on one part the click and ratchet escapement acted, being moved by the electro-magnets, while the teeth of the other part were employed to lock the train and prevent its running forward from the action of the wind on the hands. The large wheel revolved once in two hours, the spindle of which projected beyond the frame, and carried a bevelled wheel of twelve inches in diameter, placed vertically, which revolved within it. In order to give motion to the vertical rod already described, the bevelled vertical wheel worked in a second bevelled wheel placed horizontally; and above the first, on the axis of the horizontal bevelled wheel, the vertical rod or shaft revolved, and by means of wheel-work at the top of the shaft the hands of the clock were also made to revolve. The whole was kept in motion by a series of powerful electro-magnets, eight in number, on which was wound a total length of twenty-five thousand feet of copper wire, the weight of which was nearly one-and-a-half hundred. Six small batteries were used in connection with the electro-magnets. Besides the twenty-four feet dial on the south side of the transept, two smaller dials already alluded to, each of five feet in diameter, were fixed in front of the galleries, at the east and west end of the building respectively, in the centre line of the main aisle or nave. All the dials were governed by one pendulum, which was kept in motion by a new plan. The magnet was employed merely to bend a spring at each vibration to a certain fixed extent, the reaction of the spring giving the necessary impulse to the pendulum, by which means the variations which were continually taking place in the batteries had no effect on the time measured by the pendulum. At the end of each vibration of that mechanism it came in contact with a small spring tipped with platinum, which completed the necessary circuit for giving motion to the several clocks. The impulse spring was screwed on to a brass stud fixed on the bed-plate, through a slot in which the pendulum vibrated. It had a small arm extending nearly at right angles, and a second arm that projected from the armature, which being attracted down by the action of the magnet, the poles of which passed through the bed-plates, the other end of the armature came in contact with the arm projecting from the impulse-spring, and raised it so as to lock the upper end in a detent, which was screwed on to the same stud as the impulse-spring. The pendulum in the course of its vibration came in contact with the upper part of the detent, which it lifted up, thereby leaving the impulse-spring free to drop on the side of the pendulum, and follow it for a short space of its vibration, so as to give it the necessary impetus, forming what is technically called the remontoir escapement.

Samuel Warren, in his 'Lily and the Bee,' an apologue of the Crystal Palace of 1851, thus alludes to the great electric clock there :—

> "Hark! A sound! startling my soul!
> A toll profound!
> The hollow tongue of Time,
> Telling its awful flight,
> Now, to no ear save mine!
> Heard I ever here that solemn sound before!
> Or did my million fellows hear, or note?
> Now dies the sound away—
> But upwaketh, as it goes,
> Memories of ages past! The Gone!"

Byron tells us that—

> "The hollow tongue of Time
> Is a perpetual knell. Each toll
> Peals for a hope the less!"

In the South Kensington Museum is a timepiece in ormolu and enamel, the upper part of which is formed by a globe enamelled in blue. Round the base are statuettes of the Seasons in ormolu. It was manufactured by Levy, frères, et Cie, of the Rue des Fosses du Temple, Paris. Its height is eleven inches and three quarters; and its width, twelve inches. It was exhibited in the Paris Exhibition of 1855, and was purchased for 19*l.* 4*s.*

We shall now give some account of the history of the Westminster Palace clock. In March, 1844, the late Sir Charles Barry applied to Mr. B. L. Vulliamy for a specification and plan for a clock to "strike the hours on a bell of from eight to ten tons, and, if practicable, chime the quarters upon eight bells, and show the time upon four dials about 30 feet in diameter." On April 1st following, Mr. Vulliamy replied that the clock which he should propose as fit for the purpose would be very much the most powerful eight-day clock ever made in this country or in any other, but one which he should have no difficulty in executing. To fully carry out these intentions, he said he should visit many of the principal clocks at home and abroad. Shortly afterwards, however, Lord Canning, the principal Commissioner of Woods and Forests, applied to Mr. Airy, of Greenwich Observatory, for his advice upon the subject; and Mr. Airy recommended that Mr. E. J. Dent, of Cockspur Street, who was then constructing the clock for the New Royal Exchange, and Mr. John Whitehurst, of Derby, should also be applied to. This was done, and Mr. Dent, by letter dated November 14th, 1845, consented to be a competitor for the great clockmaking. Subsequently Mr. Vulliamy had cause, it seems, to retire from the competition, and Mr. Dent, having a friend in Mr. Airy, was entrusted with the order for the clock. The conditions upon which the competition was based were, in their original state, curious, and as they specially refer to the making of the largest clock known, we here append them :—

Conditions to be observed in regard to the Construction of the Clock of the New Palace of Westminster :—

I.—*Relating to the Workmanlike Construction of the Clock.*

1. The clock frame is to be of cast-iron and of ample strength. Its parts are to be firmly bolted together. Where there are broad bearing surfaces, these surfaces are to be planed.

2. The wheels are to be of hard bell-metal, with steel spindles

working in bell-metal bearings, and proper holes for oiling the bearings. The teeth of the wheels are to be cut to form on the epicycloid principle.

3. The wheels are to be so arranged than any one can be taken out without disturbing the others.

4. The pendulum pallets are to be jewelled.

II.—*Relating to the Accurate Going of the Clock.*

5. The escapement is to be dead beat, or something equally accurate, the recoil escapement being expressly excluded.

6. The pendulum is to be compensated.

7. The train of wheels is to have a remontoir action, so constructed as not to interfere with the dead-beat principle of the escapement.

8. The clock is to have a going fusee.

9. It will be considered an advantage if the external minute-hand has a discernible motion at certain definite seconds of time.

10. A spring apparatus is to be attached for accelerating the pendulum at pleasure during a few vibrations.

11. The striking machinery is to be so arranged that the first blow for each hour shall be accurate to a second of time.

III.—*Relating to the possible Galvanic Connection with Greenwich.*

12. The striking detent is to have such parts that, whenever need shall arise, one of the two following plans may be adopted (as after consultation with Mr. Wheatstone, or other competent authorities, shall be judged best), either that the warning movement may make contact, and the striking movement break contact, for a battery, or that the striking movement may produce a magneto-electric current.

13. Apparatus shall be provided which will enable the attendant to shift the connection by means of the clock action successively to different wires for different hours, in case it should hereafter be thought desirable to convey the indications of the clock to several different places.

IV.—*General Reference to the Astronomer Royal.*

14. The plans, before commencing the work, and the work when completed, are to be subjected to the approval of the Astronomer Royal.

15. In regard to the Articles 5 to 11 the maker is recommended to study the construction of the Royal Exchange clock.

These conditions bear date June 22nd, 1846, and on February 11th, 1847, in another communication to the competitors, Mr. Airy stated it was desirable that conditions 3, 7, 8, and 14, should "be fully complied with," because "the constructions to which they refer being not in extensive use in this country, they may, in the adaptation of ordinary plans to the new clock, be inadvertently passed over. But it is not intended that the stringency of the other conditions shall be in any degree relaxed." Afterwards he appended a new condition:—

16. The hour wheel is to carry a ratchet-shape wheel or a succession of cams, which will break contact with a powerful magnet, on the principle recommended by Mr. Wheatstone, at least as often as once in a minute, for the purpose of producing a magneto-electric current, which will regulate other clocks in the New Palace.

The tenders were sent in as follows:—Mr. E. J. Dent, August 8th, 1846 (conditions 1 to 15), 1,500*l.*; and March 15th, 1847 (including 16), 1,600*l.*; Mr. John Whitehurst, September 24th, 1846 (conditions 1 to 15), 3,373*l.*, and March 15th, 1847 (including 16), 3,523*l.* Mr. Vulliamy sent in designs, but he afterwards declined further to compete. Thwaites and Reid applied to the Commissioners for leave to tender, but the same was not granted. The clock now set up at Westminster was ultimately made by Mr. Dent, junior, from the designs of Mr. Denison, about 1855.

The four dials are twenty-two feet in diameter, and the figures upon them, thickly gilt, are relieved from a blue surface. These dials are considered to be the largest in the world with a minute-hand, which, on account of its great length, velocity, weight, friction, and the action of the wind upon it, requires at least twenty times more force to drive it than the hour-hand.

This clock goes for a week. The great wheel of the going part is twenty-seven inches in diameter; the pendulum is fifteen feet long, and weighs six hundred and eighty pounds; and the scape-wheel, which is driven by the musical box spring, weighs about half-an-ounce. All the wheels except the scape-wheel are of cast-iron, and all have five spokes. The barrel is twenty-three inches in diameter, but only fourteen inches long, as it does not require a rope above a quarter-of-an-inch thick. The second wheel is twelve inches in diameter. The great wheels have all one hundred and eighty teeth, the second wheel of the hour-striking part has one hundred and five, and a pinion of fifteen. The great wheels in the chiming part of the clock are thirty-eight and a-half inches in diameter. The clock is said to be at least eight times as large as a full-sized cathedral clock. It affords its keepers two hours' work a week in winding it up. It goes with a rate of under one second a week, in spite of any atmospheric changes. Respecting its present condition, Mr. Denison wrote to 'The Times' on August 11th, 1865, as follows:—"I may as well correct a mistake, which I often have to correct privately about the great clock. In consequence of the ambiguous language of another report of the Astronomer Royal, some people imagine that the clock is controlled by electric connection with Greenwich Observatory. It contains no machinery whatever for that purpose. It reports its own time to Greenwich by electrical connection, and the clockmaker who takes care of it receives Greenwich time by electricity, and sets the clock right whenever its error becomes sensible, which seldom has to be done more than once a month. Mr. Airy's last report upon the rate was 'that it may be relied on (that is, the first blow of the hour) within less than one second a week;' which is seven times greater accuracy than was required in the original conditions." On October 1st, 1859, the great bell of Westminster sounded for the last time, and while in the act of striking it became dumb for ever. Its predecessor was similarly ill-fated.

Mr. Cowper stated in 1860 that 20,300*l*. had been spent on Westminster Palace clock up to that time, and that an additional sum of 1,750*l*. was in course of being expended.

A correspondent of 'Notes and Queries' for March 23rd, 1861, relates the following account of a curious omen or coincidence:—"On Wednesday night, or rather Thursday morning, at three o'clock, the inhabitants of the metropolis were roused by repeated strokes of the new great bell at Westminster, and most persons supposed it was for a death in the Royal family. There might have been about twenty slow strokes, when it ceased. It proved, however, to be due to some derangement of the clock, for at four and five o'clock, ten or twelve strokes were struck instead of the proper number. On mentioning this in the morning to a friend, who is deep in London antiquities, he observed that there is an opinion in the city that anything the matter with St. Paul's great bell is an omen of ill to the Royal family; and he added, I hope the opinion will not extend to the Westminster bell. This was at eleven on Friday morning. I see by 'The Times' this morning, that it was not till one A.M. the lamented Duchess of Kent was considered in the least danger, and as you are aware she expired in less than twenty-four hours . . . . . . I am told the same notion obtains at Windsor." On April 1st, in the same year, at one o'clock in the dead of the night, the bells of this clock struck thirty-seven. At two and three o'clock they gave the right numbers, but at four o'clock they were again erratic, and struck thirty-nine. On both occasions when these extravagant numbers were given, the striking of the hours commenced before the quarter-chimes, these coming in towards the close.

Apropos of our mention of the great bell on which the hours are struck by the Westminster clock, we may add that from the earliest times huge bells have been used for the sounding of the hours. For nearly three centuries has the number of them been struck upon the bell called Great Tom, at Christ's College, Oxford, by a heavy clock-hammer, weighing fifty-four pounds and a half. This bell is five feet nine inches in height, twenty-one feet in girth, and seven tons fifteen hundredweights, or rather over seventeen thousand pounds in weight, without the clapper, which weighs three hundred and forty-two pounds. The College clock-bell at Canterbury weighs three tons and ten hundredweights; the College clock-bell at Gloucester weighs three tons and five hundredweights; and the Minster clock-bell at Beverley, in Yorkshire, weighs two tons and ten hundredweights. The north tower of the Temple, formerly the Cathedral church of St. Peter, at Geneva, contains the great bell called Clemence, which is twenty feet in circumference round the base. A single blow struck on this bell announces daily the precise instant of noon, which is determined by a sundial placed on the western front of the Temple.

In 1858, a mechanic in Bohemia invented a musical clock-bed, which was so constructed that, by means of hidden mechanism, a pressure upon the bed caused a soft and gentle air of Auber's to be played, which continued long enough to lull the most wakeful to sleep. At the head was a clock, the hand of which being placed at the hour that the sleeper wished to rise, when the time arrived the bed played a march of Spontoni's with drums and cymbals, and musical thunder enough to rouse the seven sleepers.

In 1858, a watchmaker named Chenball, of Drake Street, Plymouth, exhibited in his shop-window a clock of the size of an ordinary eight-day clock, with a novel and very simple movement, which was said to be capable of going as long as the durability of the materials permitted, without the aid of weight or spring, and in short without any manual assistance whatever.

In 1859, a locksmith in Frankfort-on-the-Maine constructed a strong box without any key-hole at all, and which even the owner himself could not open. Inside was a clock-work, the hand of which, when the box was open, the owner placed at the hour and minute when he again wanted to have access to the interior of the box. The clockwork began to move as soon as the lid was shut, and opened the lock from the inside at the moment which the hand indicated. Time, dependent upon the owner, was the key to this clock-lock—a key that could neither be stolen nor imitated.

In 1859, after years of mechanical labour, James White, of Wickham Market, completed a self-winding clock, which determined the time with unfailing accuracy, continuing a constant motion by itself, never requiring to be wound up, and being capable of perpetuating its movements so long as its component parts should exist.

During the war in China in 1860, the Emperor's Summer Palace near Pekin was sacked by the allied forces, and much treasure was taken therefrom. Among the loot were many clocks and jewelled watches. A French soldier took a figure of a camel in solid silver, nearly twenty inches high, bearing on its back a clock, and its hump being decorated with rubies, emeralds, and other precious stones. The statuette, on which the word "London" was engraved, came into the possession of a non-commissioned officer of the 101st Regiment, who refused seventy thousand francs for it. Lord Amherst's watch, worth 200*l.*, was seized, and sold by a French soldier for twenty dollars.

In the International Exhibition of 1862, in the middle avenue, was a clock by Benson, which struck the hours and quarters on five bells; the largest weighing twenty-two hundredweight. The works were three hundred feet from the dial, which was situated in the great central tower, the connections being carried underground. The weights exceeded a ton, and were two hundred feet from the works. A new remontoir escapement allowed the use of the great weight required to drive a clock of such size, so distant from the dial, and moving hands of such magnitude, sometimes in opposition to the wind. A two-seconds compensation-pendulum was employed. In the south-east transept was a turret-clock by Dent, which struck the hours on a bell weighing between three and four tons, and the quarters on four smaller bells. The wheels were of gun-metal; and each of the four dials was seven feet in diameter. The Exhibition also contained numerous other horological machines, among which we may mention a steam or speed clock, a chime-clock with fifty changes, silent clocks, cuckoo-clocks, a clock with a perpetual register of the day of the week and month, an astronomical clock impelled by gravitation, a regulator to be wound up once in twelve months, and a geographical clock showing the time throughout the world.

In various churches in England it has been found inconvenient to have the dial of the clock outside the building, either because there has not been room enough for it, or because it would be a disfigurement to the architecture. In such cases striking clocks have been placed, which sound forth the hours and quarters, although not indicating them visibly. The only church-clock in London without a face is St. Vedast's, in Foster Lane, at the back of the Post Office. It strikes on a small shrill bell. Both Litchfield and Peterborough Cathedrals have these faceless clocks. There is now at Leeds Castle, Kent, a clock of the early part of the sixteenth century, which has the movement and striking-part complete, but no dial-works or face. A butcher-churchwarden of Doncaster ruined the grand old tower of the church there by placing a hideous clock-face in it, which was so constructed that no one could see the time by it except from the butcher's own door.

The wooden-clocks, commonly known as Dutch clocks, have been made for about two centuries. Before the cheap American clocks were introduced there was scarcely a middle-class and lower-class householder that did not possess a Dutch clock, which told the time in the kitchen or bedroom, where its painted face and cords and weights were mostly to be found, with tolerable accuracy. But we remember what Albert Smith said in his 'Christopher Tadpole' about this humble horologe: "The Dutch-clock pointed to twenty minutes to three and struck eleven—the combination signifying that it was eight precisely, after the dissolute manner of Dutch clocks in general." These clocks were first produced in Holland; but the greater number of them are now made on the confines of the Black Forest, in Germany. This Forest is the mountainous district, extending from the frontier of Switzerland northward along the Rhine, a distance of about one hundred miles. It contains an abundance of wood, almost its sole production, which is exported in large quantities to England and Holland. The peasantry, an intelligent and frugal class, make good use of this natural produce, in the skilful manufacture of various useful and ornamental articles. But their greatest mechanical genius is expended upon their clocks. As a rule, these branches of industry are carried on in the homes of the people after the agricultural work is done. It is during the long winter evenings especially that they, both old and young, employ their time in these occupations. The labour is divided among the case-makers, the founders of the brass wheels and bells, the chain and chain-wheel makers, the painters and varnishers, and, lastly, the clockmakers, who put the works together and finish them. With the exception of the works, which are cast by the founder, and the painting of the dial, the whole of the clock is made by the members of one family. At an early age the children begin to carve the roughest woodwork, and advance through the other stages of the labour, until they become clockmakers and finishers. In and after the sixteenth century there was a gradually increasing demand for clocks in Germany; but they were too high in price to be generally used in the house, and for many years the makers endeavoured to produce a cheaper kind. In 1660 some intelligent wood carvers in the Black Forest succeeded in making clocks entirely of wood, which, although rough and simple, for the time supplied the want. But they showed only the hours, and would go only for half-a-day. There was no striking work, and the moving power was a balancing piece of wood with two movable weights. In 1740, instead of the balancing piece a pendulum was used as a motive principle, and the improvements in a short time made the clocks to go twenty-four hours, and strike the hours and quarters. Some also moved figures, indicating the day of the month, and the like. In 1750, the wooden works were replaced by metallic works, wheels, and chains. To meet the increasing competition which improved machinery caused in the clock trade, the Grand Duke Leopold, about 1847, founded the Clock and Watch Makers' School at Furtwangen, which is supported by the Government of Baden with a yearly contribution of 1,000*l*. The annual exports of clocks from the Grand Duchy of Baden alone, not including watches, amounts to one million pounds sterling.

In a building in Yarmouth up to the year 1861 was an ancient timepiece called the Dutch clock; the building in which it was placed having been used in the seventeenth century as a chapel for the Dutch refugees.

Apropos of our mention of Dutch clocks, we may add that clock-chimes or carillons were invented in the Low Countries, where they have been brought to the greatest perfection, and may be heard in every town. They are of two kinds, the one being attached to a cylinder, like the barrel of an organ, which always repeats the same tunes, and is moved by machinery; and the other being a superior kind, played by a musician with a set of keys. In all the great towns are amateurs or salaried professors, usually the organists of the churches, who perform with great skill upon this gigantic instrument, which is placed high up in the steeple. So fond are the Dutch and Belgians of this kind of music, that in some of their cities the chimes appear scarcely ever to be at rest, either by day or by night. The tunes are usually changed every year. Chimes were in existence at Bruges in 1300; therefore the claim of Alost to the invention in 1487 is unfounded. The chimes from the Tour des Halles, Bruges, are to English ears most novel and pleasing, although not to be compared with the sprinkling shower of music which drops down from the high tower of Antwerp Cathedral every few minutes of the day and night. Longfellow thus wrote of the chimes of Bruges:—

> "But amid my broken slumbers,
> Still I heard those magic numbers,
> As they loud proclaimed the flight
> And stolen marches of the night;
> Till their chimes in sweet collision
> Mingled with each wandering vision,
> Mingled with the fortune telling
> Gipsy-bands of dreams and fancies,
> Which amid the waste expanses
> Of the silent land of trances
> Have their solitary dwelling.
>
> "Thus dreamed I, as by night I lay
> Listening with a wild delight
> To the chimes that, through the night,
> Rang their changes from the belfry
> Of that quaint old Flemish city."

The chimes of Antwerp Cathedral consist of nearly one hundred bells; and a singular feeling results from them when they play under the feet of a person on the summit of the tower (on which is the gigantic clockface). They throb through the very stones and frame of the building, and every nerve of the body.

Longfellow, in his exquisite prose poem, 'Hyperion,' gives the following solemnly impressive description of midnight bells in the quiet city of Heidelberg: "In conversation like this, the hours glided away, till at length, from the Giant's Tours, the castle-clock struck twelve, with a sound that seemed to come from the Middle Ages. Like watchmen from the belfries, the city clocks answered it, one by one. Then distant and muffled sounds were heard. Inarticulate words seemed to blot the foggy air, as if written on wet paper. These were the bells of Handschuhsheim, and of other villages on the broad plain of the Rhine, and among the hills of the Odenwald,—mysterious sounds, that seemed not of this world."

The public clocks of Belgium strike the hour half-an-hour beforehand; thus at half-past eleven they strike twelve. We are at a loss to understand the meaning or utility of this arrangement.

Mr. Vulliamy says, "There are several public clocks in Flanders which have very large dials; but these only indicate the hours and the quarters by one hand, which makes a revolution in twelve hours: indeed, this was the only mode by which, until very recently, the time was shown by public clocks. These dials are not inaptly called skeleton dials, being composed of very light circles connected together by the figures and other divisions and slender bars traversing the centre, and are fixed in their places by fastenings not perceptible at a little distance. By this means the sight of the building is very little interrupted, or its general appearance disfigured. It must be borne in mind that when these buildings were erected clocks were almost unknown, and that not any provision being made for a dial, the general appearance is much disfigured by a large dial as commonly made. Of this there cannot be a more striking example than is exhibited by the clock-tower of Canterbury Cathedral, which is sadly disfigured by the clock-face. In modern buildings, when it is intended to place a clock, it is customary to provide for the dial. The largest skeleton dial in Flanders is the dial of the clock in the great tower of the Cathedral at Malines; it is forty-two Flemish feet* in diameter, but from its great height, and the great size of the tower, the sides of which measure ninety feet at its base, it does not appear nearly so large." Mr. Vulliamy also adds: "I will here notice that the clocks in Belgium strike upon bells comparatively small. These circumstances very much diminish the difficulties attendant upon making a clock of this description; moreover, the performance of most of the Flemish clocks is such as would not be tolerated in the situation for which the clock for the new Palace (at Westminster) is required."

American clocks have found great favour with the public, and by reason of their portability and the neatness of their exterior have much superseded the old familiar Dutch clocks. They are of inferior workmanship, and lack altogether that finish of which the English workman is justly proud. America has organized a very extensive system of clock manufacture, which is carried out on the factory system, chiefly at Connecticut. At that place one clockmaking firm employs two hundred and fifty hands. Many of the operatives are boys and girls, and the products of their united labours are six hundred clocks a-day. 'Simmonds's Colonial Magazine' for 1845 contains an account of the clock-factory of Jerome, in the city of New Haven, then one of the most extensive establishments of the kind in the United States. The writer says: "We cannot describe minutely the whole process of making a clock, or the life-like movement of the machinery; it would take more time and space than we can at present devote to this purpose. In short, the case, movements, plates,

* A Flemish foot is equal to about 12·37 inches English measurement, which gives 43 English feet 2·7 inches for the diameter of the dial.

face, &c., which when put together, form one of Jerome's celebrated 'brass eight-day clocks,' go through some fifty different hands before completed. One man can put together about seventy-five movements per day, while every part, from the first process to the finishing, goes on with equal rapidity. We learn from him that the greatest bulk of clocks which he anticipates making this year are designed for European markets, and that he has already received orders from houses in London and Birmingham, England, a large house in Scotland, and also some quite extensive dealers in Canada. In fact, the Yankee clock is becoming a general favourite in England, almost entirely superseding the old Dutch clock, which has been long used there as a time-piece. He yearly consumes of the various articles used in the manufacture of clocks the following enormous quantities:—500,000 feet pine lumber; 200,000 feet mahogany and rosewood veneers; 200 tons of iron for weights; 100,000lbs. of brass; 300 casks of nails; 1,500 boxes of glass, 50 feet per box; 1,500 gallons varnish; 15,000lbs. wire; 10,000lbs. glue; 30,000 looking-glass plates. 2,400 dollars are paid yearly for printing labels, and for screws, saws, coal, and oil. Workmen employed, 75; paid wages yearly, 30,000 dollars; clocks made per day, 200; year, 50,000."

The wheels and plate-holes of American clocks are all stamped; in fact there is very little worth of manual labour in the whole of their movements. The pinions are all of a kind that are called lantern pinions, which have their leaves made of pieces of wire set round an axis in two collars. A traveller in the Mosquito territory in Central America, writing in 1856, of a visit paid by him to the negro king of the country, says that a Yankee clock was part of the furniture of his state room. That which is accomplished in an American clock by a spring, the going, was in the tall old-fashioned eight-day clocks performed by the gradual fall of a heavy weight. While writing of these domestic clocks of our forefathers, we are reminded of the lines of Longfellow upon—

"THE OLD CLOCK ON THE STAIRS.

"Somewhat back from the village street
Stands the old-fashioned country seat.
Across its antique portico
Tall poplar-trees their shadows throw;
And from its station in the hall
An ancient timepiece says to all,—
'For ever—never!
Never—for ever!'

"Halfway up the stairs it stands,
And points and beckons with its hands
From its case of massive oak,
Like a monk who, under his cloak,
Crosses himself, and sighs, alas!
With sorrowful voice to all who pass,—
'For ever—never!
Never—for ever!'

"By day its voice is low and light;
But in the silent dead of night,
Distinct as a passing footstep's fall,
It echoes along the vacant hall,
Along the ceiling, along the floor,
And seems to say, at each chamber-door,—
'For ever—never;
Never—for ever!'

"Through days of sorrow and of mirth,
Through days of death and days of birth,
Through ever swift vicissitude
Of changeful time, unchanged it has stood,
And as if, like God, it all things saw,
It calmly repeats those words of awe,—
'For ever—never!
Never—for ever!'

CLOCK ON THE STAIRS.

"In that mansion used to be
Free-hearted hospitality;
His great fires up the chimney roared;
The stranger feasted at his board;
But, like the skeleton at the feast,
That warning timepiece never ceased,—
'For ever—never!
Never—for ever!'

"There groups of merry children played,
There youths and maidens dreaming strayed;
O precious hours! O golden prime,
And affluence of love and time!
Even as a miser counts his gold,
Those hours the ancient timepiece told,—
'For ever—never!
Never—for ever!'

"From that chamber, clothed in white,
The bride came forth on her wedding night;
There, in that silent room below,
The dead lay in his shroud of snow;
And in the hush that followed the prayer,
Was heard the old clock on the stair,—
'For ever—never!
Never—for ever!"

"All are scattered now and fled,
Some are married, some are dead;
And when I ask, with throbs of pain,
'Ah! when shall they all meet again?'
As in the days long since gone by,
The ancient timepiece makes reply,—
'For ever—never!
Never—for ever!'

"Never here, for ever there,
Where all parting, pain, and care,
And death and time shall disappear,—
For ever there, but never here!
The horologe of Eternity
Sayeth this incessantly,—
'For ever—never!
Never—for ever!'"

Longfellow puts as a motto to this poem the following extract from Jacques Bridaine:—"L'éternité est une pendule, dont le balancier dit et redit sans cesse ces deux mots seulement, dans le silence des tombeaux: 'Toujours! jamais! Jamais! toujours!'" There is no dead thing so like a living thing as a clock, which deliberately performs its appointed work both by day and by night, with scarcely any interruption during the lapses of many generations of men, reminding them that they are passing away, and of the period when "the great clock of Time will have run down for ever."

France is famous for highly ornamental table and mantelshelf clocks, the cases of which are mostly of more worth than the works. These articles have long formed an important feature in the decoration of rooms in French households. A French traveller, writing in 1809, tells us that luxury did not extend to the interior of the houses in Spain, for he saw there none of those glasses and clocks which embellished the apartments of his native country. A magazine writer, in 1856, referring to a Parisian home, says:—"The strange little rooms looked even stranger by day; nothing abounded but looking-glasses and clocks, which were hung on every vacant place."

The comparatively recent mission of the Earl of Elgin to China and Japan (1857–1859) has brought under the notice of Europeans the time measurers in use among the people of the latter country. Mr. Oliphant, in his 'Narrative' of the mission says:—"When we got back into the business part of the town (Yedo), we stopped at a watchmaker's to buy jewellery and clocks . . . . The latter were of various descriptions, some constructed on European models, others fashioned upon a principle peculiar to Japan, and supposed to be more convenient for the registration of the singular division of their time. The twenty-four hours are divided in Japan into twelve periods of time, six of which are appropriated to darkness, and six to the light. The day being calculated from sunrise to sunset, there is a necessary variation in the length of the six day and six night hours, the latter being the longest in winter, the former in summer. The clocks are altered periodically to suit the seasons of the year."

On a similar principle to that of the primitive clepsydra, or water-clock, is the atmospheric clock, an invention which was patented a few years ago; the main difference being in the density of the fluid used, mercury in the present machine being substituted for water. This clock is in appearance like a long thermometer, without the bulb of mercury at the bottom. It has a glass tube, about three-eighths of an inch in diameter, and the length of the thermometer-like frame; this tube is secured to the frame by two bands, through which the tube easily slides. Inside the glass tube is another and smaller tube, at each end of which is some mercury and a scrap of blotting paper, or other absorbent material, for the purpose of absorbing any damp which might find its way into the tube. About an inch and a-half from each end of the inner tube is a small throat, through which the mercury has to pass. On each side of the glass tube are the divisions of time. The mercury in the top end of the tube is placed opposite the mark of the proper time, and it descends to the bottom of the tube exactly as the time lapses. When the mercury has reached the bottom of the tube the frame can be turned, and the mercury set to the time on the other side; and so the time may be continually indicated. This is a sort of perpetual hour-glass.

Some time since was exhibited at a watchmaker's window in Montgomery Street, San Francisco, a clock, the hand of which apparently could not be made to revolve, there being no visible means to make it do so. The dial was a simple plate of transparent glass, with a small, smooth pin in the centre, which passed through a plain hole in the hand, there being only one, an hour-hand, light and slender; and upon the short end of it was a small box of thin metal. There was no contact of the hand with the dial except at the pivot, and there was nothing touching the pivot except the hand and the glass in which it was fixed; yet the machine kept perfect time. This clock was a mechanical puzzle, and attracted a constant crowd round the window in which it was displayed. Probably the works were in the small box on the short end of the hand. Suppose within this box watchwork driven by a spring and regulated by a balance-wheel, so that it would cause a little hand to revolve once in twelve hours, in a plane parallel with the dial; then let this little hand carry a small weight, say a pistol bullet, on its end, and let the large hand be made very light and be so nicely poised that when the weight was furthest from the fulcrum, it would bring the short end of the large hand down, causing the long end to point directly upward and to indicate twelve o'clock; but when the weight was nearest the fulcrum the long hand would overbalance, and so point downward to six o'clock. Then as the weight revolved it would cause the hand to balance in the several positions around the dial, depending upon the time of day as kept by the watchwork within the box. If this explanation be correct, the hand might be laid away in a drawer, and on taking it out at any time, and slipping it upon the pivot, it would swing to the exact hour of the day.

The total number of clocks imported to this country in 1865, to August 31st, was one hundred and forty-nine thousand nine hundred and sixteen, as against two hundred and twelve thousand three hundred and fifteen in 1864, and one hundred and ninety-six thousand and eighty-seven in 1863, corresponding periods.

The following proverbs and items of folk lore relate to clocks:—"Lovers ever run before the clock;" "The clock goes as it pleases the clerk;" "Whores and thieves go by the clock;" "A mill, a clock, and a woman always want mending;" "They agree like the clocks of London." This last proverb is found among both the French and Italian ones, for an instance of disagreement. There is a Somersetshire saying, that "A child born in chime-hours will have the power to see spirits." A Staffordshire proverb says, "You're too fast, like Walsall clock." It is said in Peterborough, that when the clock of the cathedral and the clock of the parish church strike simultaneously, there will be a death in the Minster Yard. This is fully believed by the old women of the city; although the saying probably arose from the noted irregularities of the two clocks. A writer in 'Notes and Queries' says:—"I knew an intelligent, well-informed gentleman in Scotland, who, among the last injunctions on his death-bed, ordered that as soon as he expired the house-clock was to be stopped, which was strictly obeyed. His reason for this I never could fathom, except that it was to impress upon his family the solemnity of the circumstance, and that with him time was no longer." The custom of stopping clocks in a house wherein a corpse lies prevails in Germany. It is simply a manifestation of respect for the dead, and is designed to produce that complete silence which is supposed to be the most proper token of respect.

The clock of Trinity College, Dublin, is always kept a quarter of an hour slow, and all University examinations and proceedings are regulated by that time. A legend states that the College time was altered in consequence of a student being killed in endeavouring to cross the railings, having been too late for admission by the gate. A traveller about the year 1730 tells us that the clocks at Basil went an hour faster than in other places, which some derived from the discovery of a conspiracy, the measures whereof were defeated by the alteration of the clock. Others, from the time of the council held there, which after lasting seventeen years ended in 1448, as a contrivance that the holy fathers should rise an hour earlier, or sit at table an hour less, two o'clock being the time of the council's sitting.

Two persons belonging to a neighbouring town, being on a visit at Glasgow to see the lions, went to the College among other places. On looking up to the clock-dial they were astonished to observe only one hand, which was an hour behind. One of them thinking that nothing could be wrong about the College, observed in a flippant, apologetic tone, "Hoot man, that's naething ova; 'od, man, I've seen our toun clock aught days wrang." Yorkshire men always speak of a clock as "she."

Mr. J. C. Jeaffreson in his novel, 'Live it Down,' 1863, gives us the following legend about Blownorton clock:—"There was a pleasant fiction throughout the 'lights lands,' that certain inhabitants of the obscure parish of Blownorton had in their keeping a magnificent eight-day clock, which they held on trust to present it to the first man in the 'light lands' who could prove that he habitually minded his own business, and never meddled intrusively with the affairs of his neighbours. There were many firm believers in the existence of this chronometer, though no one could name the eccentric philanthropist who in the mists of antiquity, had purchased and bequeathed it to the Blownorton trustees, for the purpose of inciting mankind to virtue and prudence. The fiction did not lose in humour from the fact that, at the close of the last century, Blownorton was described by a grave photographer of the 'light lands' as a decayed parish, containing only sixteen souls, since which time the said population of sixteen souls had so decreased from sickness, emigration, and the destruction by fire of its two last cottages, that it was generally held in popular imagination to have become an altogether uninhabited tract of thirsty sand,—whereon nothing was to be found save the aforementioned costly timepiece, which no one had ever proved himself worthy of possessing, and which could not any longer be disposed of according to the terms of the trust, as all the inhabitants of Blownorton had vanished: and no person or persons existed in all creation qualified to bestow the clock on any fit candidate who should lay claim to the prize. It is to be hoped that the Charitable Trust Commissioners have made enquiries concerning this clock. As gentlemen whose especial business it is to look into other people's business, the right of any one of the said Commissioners to take possession of the clock, in consideration of his own individual merits, would be an interesting subject for discussion."

Dr. Bigsby tells the author that he once saw at a

furniture broker's shop in the country a very old and curious looking timepiece exposed for sale. He asked the dealer if it were Dutch or English. "Oh, English," said the owner; "the maker's name is on the dial, and I have often seen clocks of his make." "What is the name?" asked the doctor. "Tummas Fudgit," was the reply. The doctor was puzzled at the moment, but on examining the dial, upon which certain words indistinctly appeared, engraved in very corroded steel, he read the often-repeated warning — "Tempus fugit," which the shopkeeper had so ridiculously misunderstood. This story reminds us of another one told of an auctioneer at Nottingham, who, enlarging upon the excellence of an engraving, said that nothing more need be added to show the value of the piece, than to mention that it was by that most eminent and well-known artist, Pinxit.

Lightning Source UK Ltd.
Milton Keynes UK
UKOW040146131112

202093UK00001B/30/P